Lecture Notes in Networks and Systems

Volume 30

Series editor

Janusz Kacprzyk, Polish Academy of Sciences, Warsaw, Poland
e-mail: kacprzyk@ibspan.waw.pl

The series "Lecture Notes in Networks and Systems" publishes the latest developments in Networks and Systems—quickly, informally and with high quality. Original research reported in proceedings and post-proceedings represents the core of LNNS.

Volumes published in LNNS embrace all aspects and subfields of, as well as new challenges in, Networks and Systems.

The series contains proceedings and edited volumes in systems and networks, spanning the areas of Cyber-Physical Systems, Autonomous Systems, Sensor Networks, Control Systems, Energy Systems, Automotive Systems, Biological Systems, Vehicular Networking and Connected Vehicles, Aerospace Systems, Automation, Manufacturing, Smart Grids, Nonlinear Systems, Power Systems, Robotics, Social Systems, Economic Systems and other. Of particular value to both the contributors and the readership are the short publication timeframe and the world-wide distribution and exposure which enable both a wide and rapid dissemination of research output.

The series covers the theory, applications, and perspectives on the state of the art and future developments relevant to systems and networks, decision making, control, complex processes and related areas, as embedded in the fields of interdisciplinary and applied sciences, engineering, computer science, physics, economics, social, and life sciences, as well as the paradigms and methodologies behind them.

Advisory Board

More information about this series at http://www.springer.com/series/15179

Rahamatullah Khondoker
Editor

SDN and NFV Security

Security Analysis of Software-Defined
Networking and Network Function
Virtualization

 Springer

Editor
Rahamatullah Khondoker
Mobile Networks (MNE)
Fraunhofer Institute for Secure Information
   Technology
Darmstadt
Germany

ISSN 2367-3370   ISSN 2367-3389   (electronic)
Lecture Notes in Networks and Systems
ISBN 978-3-319-71760-9   ISBN 978-3-319-71761-6   (eBook)
https://doi.org/10.1007/978-3-319-71761-6

Library of Congress Control Number: 2017959899

Printed on acid-free paper

This Springer imprint is published by Springer Nature
The registered company is Springer International Publishing AG
The registered company address is: Gewerbestrasse 11, 6330 Cham, Switzerland

*To my master thesis supervisor
Prof. Dr. Carmelita Görg who was a
Professor at the Department of
Communication Networks in the University
of Bremen, Germany.*

Preface

This textbook is intended for IT security professionals, engineers, and researchers who need IT security recommendations for deploying SDN and NFV technologies. The recommended security solutions should be taken as suggestions and must be investigated beforehand before deploying those solutions in the real world. The authors, editors, and publishers of this book will not take any responsibility if any harm happens due to using the suggested security solutions.

Darmstadt, Germany   Rahamatullah Khondoker
November 2017

Acknowledgements

At first, I would like to acknowledge the contribution of Saeed Ehteshamifar and Marco Braeuning for helping me in formatting the book. This book cannot be composed without the individual contributions of the authors. I would like to thank them as well. Lastly, I have to thank Dr. Thomas Ditzinger and Ms. Varsha Prabakaran, Springer to publish this book.

Contents

Contributors

Parvez Ahmad Department of Computer Science, TU Darmstadt, Darmstadt, Germany

David Artmann Department of Computer Science, TU Darmstadt, Darmstadt, Germany

Marco Bräuning Department of Computer Science, TU Darmstadt, Germany

Ankush Chikhale Department of Computer Science, TU, Darmstadt, Germany

Qamar Ilyas Department of Computer Science, TU Darmstadt, Darmstadt, Germany

Sven Jacob Department of Computer Science, TU Darmstadt, Darmstadt, Germany

Rajat Jain Department of Computer Science, TU Darmstadt, Darmstadt, Germany

Rahamatullah Khondoker Fraunhofer SIT, Darmstadt, Germany

Timm Lippert Department of Computer Science, TU Darmstadt, Darmstadt, Germany

Anagha Anilkumar Sagare Department of Computer Science, TU Darmstadt, Darmstadt, Germany

Acronyms

AAA	Authentication, Authorization, and Accounting
ACL	Access Control List
AH	Authentication Header
AP	Access Point
API	Application Programming Interface
APIC	Application Policy Infrastructure Controller
ARP	Address Resolution Protocol
AS	Authentication Service
ASP	Authentication Service Provider
AVC	Application Visibility and Control
BGP	Border Gateway Protocol
BSSID	Basic Service Set Identification
CM	Cloud Monitor
CMS	Cloud Management Service
DCN	Data Center Network
DDoS	Distributed Denial of Service
DDS	Data Distribution Service
DFD	Data Flow Diagram
DHCP	Dynamic Host Configuration Protocol
DMVPN	Dynamic Multipoint Virtual Private Network
DNS	Domain Name System
DoS	Denial of Service
DPI	Deep Packet Inspection
DRES	Distributed Real-Time and Embedded System
DTP	Dynamic Trunking Protocol
EM	Enterprise Module
GNV	Global Network View
GRM	Global Resource Manager
HDFS	Hadoop Distributed File System
HTTP	Hypertext Transfer Protocol

HTTPS	Hypertext Transfer Protocol Secure
IDS	Intrusion Detection System
IKE	Internet Key Exchange
IoT	Internet of Things
IP	Internet Protocol
IPS	Intrusion Prevention System
JSON	JavaScript Object Notation
JVM	Java Virtual Machine
LVAP	Light Virtual Access Points
MAC	Mandatory Access Control
MAC	Message Authentication Code
MOF	Map Output File
MPLS	Multiple Protocol Label Switching
MU-MIMO	Multiple-User Multiple-Input and Multiple-Output
NBAR	Network-Based Application Recognition
NBI	Northbound Interface
NCS	Network Control System
NIC	Network Interface Card
NP	Network Provisioner
NRS	Network Runtime Status
OS	Operating System
OSGP	Open Smart Grid Protocol
PFR	Performance Routing
PKI	Public Key Infrastructure
QDR	Quad Data Rate
QoS	Quality of Service
RADIUS	Remote Authentication Dial-In User Service
RBAC	Role-Based Access Control
RDMA	Remote Direct Memory Access
REST	Representational State Transfer
RNIC	RDMA Network Interface Card
SA	Switch Adapter
SASL	Simple Authentication and Security Layer
SBI	Southbound Interface
SDN	Software-Defined Networking
SHA	Secure Hash Algorithm
SIEM	Security Information and Event Management
SSID	Service Set Identification
SSL	Secure Sockets Layer
TCP	Transmission Control Protocol
TLS	Transport Layer Security
UDP	User Datagram Protocol
ULP	Universal Logging Protocol
VAN	Virtual Application Network
VCN	Virtual Cloud Networking

VLAN	Virtual Local Area Network
VM	Virtual Machine
WAAS	Wide Area Application Service
WAN	Wide Area Network
WLAN	Wireless Local Area Network
XML	Extensible Markup Language
XMPP	Extensible Messaging and Presence Protocol

Chapter 1
Security Analysis of SDN Routing Applications

Anagha Anilkumar Sagare and Rahamatullah Khondoker

Abstract With the steady increase in the information and high network resource sharing, organizations require big data centers. To control the workload in the data centers and minimize the response time, effective load-balancing systems are necessary. The routing applications play an important role here. Some routing applications based on Software Defined Networking (SDN) like Plug-n-Serve, Hedera, ElasticTree suggest an efficient way to handle such a traffic load in the data centers. Centralised routing makes it possible to adjust the network elements like switches, ports, links dynamically as per the traffic load. The routing application takes control of data flow management in the data center system, finds a non-conflicting way for the flow and instructs the switches accordingly. Security of routing applications is important. If an attacker takes control over the data flow routing or scheduling, it can result in forwarding traffic to the servers/switches which are controlled by the attackers. The attacker can even shut down the data center system as some data centers may rely totally on routing application for data flow management. In this paper, several SDN routing applications are compared and detail analysis of two applications Plug-n-Serve and ElasticTree are performed. The architecture of these applications is explained and the security analysis is done using a threat analysis tool called STRIDE. We suggest some mitigation techniques for the well known threats like spoofing, tampering, repudiation etc. and also check if the application has an in-built countermeasure against these threats. In this paper, we describe how ElasticTree application by design provides some mitigation techniques against the threats and the mitigation techniques that the Plug-n-Serve application could use to avoid the threats.

Keywords SDN routing applications · Plug-n-Serve · ElasticTree · STRIDE · Threat analysis methods · Plug-n-Serve DFD · Plug-n-Serve analysis · ElasticTree DFD · ElasticTree analysis

A. A. Sagare (✉)
Department of Computer Science, TU Darmstadt, Darmstadt, Germany
e-mail: anagha.sagare@gmail.com

R. Khondoker
Fraunhofer SIT, Darmstadt, Germany
e-mail: rahamatullah.khondoker@sit.fraunhofer.de; r.khondoker@yahoo.com

© Springer International Publishing AG 2018
R. Khondoker (ed.), *SDN and NFV Security*, Lecture Notes in Networks and Systems 30, https://doi.org/10.1007/978-3-319-71761-6_1

1.1 Introduction

Computing has been evolved greatly in the last years. Former communication networks (for example, 1G/2G/3G mobile networks, industrial networks, enterprise networks) used specialized and thus costly equipments which make it expensive/difficult to make an experiment at a large scale. In addition, traditionally the data and control plane elements were bounded in one network element (switch/router) and vendor-specific. Such a closed structure of the system enables the access only to the vendors, to modify the network according to customer requirements. SDN overcomes all of these hurdles and offers an approach where the control plane and the data plane are separated providing a central control point (called controller) for centrally coordinating and managing a network. The communication between the control plane and the data plane is done using a south bound application programming interface (API) protocol such as the OpenFlow protocol [1]. Forwarding of a data packet is accomplished by a forwarding device (switch/router) located in the data plane and the controller located in the control plane is responsible for taking a decision about how the packet should be routed in the network and push this decision to the forwarding device. Figure 1.1 shows the components and interfaces of the SDN architecture. The controller provides the topology information to applications from which they build an abstract view of the network. Applications use APIs to communicate with the controller. The architecture has several interfaces, a northbound interface is an API between applications and the controller (NBI API), whereas a southbound interface is an API between the controller and forwarding devices (SBI API).

An attacker could compromise the controller or could manipulate the control messages between the forwarding device and the controller. For example, a Denial of Service (DoS) attack is possible by flooding the controller-switch communication or the flow tables of a switch. To be able to apply appropriate mitigation mechanisms, a security analysis of SDN application, especially routing application is necessary.

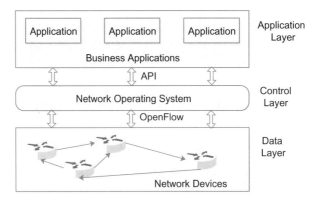

Fig. 1.1 SDN architecture

Now-a-days, many organizations have their own data centers. These data centers have high variability of workload which is unknown at the initial stage. This results in lower throughput or higher latency. To increase the throughput and decrease latency, the workload should be handled dynamically which can be done using SDN. Routing applications are used in the SDN-based data centers and enterprises. These applications obtain information about entire network topology, server statistics and (re)route the data flow accordingly. Entire routing and scheduling management are done using these routing applications which make those applications as crucial parts of the data center system. If the attacker takes control over the routing applications, then he can perform different attacks like DoS, Tampering, and Information Disclosure etc. If an attacker successfully performs any of these attacks, then an entire data center may crash. The security of several SDN routing applications is analyzed in this paper.

Some SDN-based routing applications are Plug-n-Serve, Hedera, ElasticTree, Aster*x and HyderabadApp. Out of which Plug-n-Serve and ElasticTree are chosen for security analysis using STRIDE. The paper is organised as follows: Sect. 1.2 describes the SDN routing applications and the reasons are given why two of them are chosen here for the analysis. Section 1.3 focuses on different security analysis methods and the reason for choosing STRIDE as the security analysis method. The detail security analysis of both applications using STRIDE is included in Sect. 1.4. Section 1.5 briefs future work and concludes the paper.

1.2 SDN Routing Applications

The SDN routing applications help data centers in reducing energy consumption and operating cost. Several SDN routing applications have been proposed, for example, Plug-n-Serve [2], ElasticTree [3], Aster*x [4], Hedera [5] and HyderabadApp [6]. A summary of these applications can be found in Table 1.1. The Plug-n-Serve and ElasticTree application are chosen for the security analysis. Plug-n-Serve and ElasticTree are claimed to be good for performance and energy saving. Aster*x is based on Plug-n-Serve application. The only difference is, it uses different load balancing algorithms. As the architecture is similar, performing a security analysis of the same architecture is not required. Hedera is a dynamic flow scheduling application. Architecture wise, it is similar to the Plug-n-Serve application. The architecture of HyderabadApp is based on ElasticTree architecture and also it makes the network unstable sometimes as mentioned in [6]. Hence we do not consider HyderabadApp for further analysis.

1.2.1 Plug-n-Serve

The Plug-n-Serve is an SDN-based load balancing application for data centers which uses OpenFlow as a SBI API protocol. The operators are able to increase the

Table 1.1 SDN based routing applications

Applications		
Name	Description	Technologies used
Plug-n-Serve	OpenFlow enabled load balancing application	NOX Controller, OpenFlow 0.8, Standford University Testbed
ElasticTree	Handles dynamic workload changes in the data center	NetFlow, SNMP, OpenFlow, own Testbed
Aster*x	Based on Plug-n-Serve and path selection	NOX controller, OpenFlow 1.0
Hedera	Dynamic flow scheduling system	OpenFlow controller, PortLand Testbed
HyderabadApp	Architecture similar to ElasticTree, differs in incrementally power on/off element strategy	Floodlight controller, Mininet Simulator

performance and capacity of applications (for example, web services) by simply adding computing resources and switches. It detects the addition and removal of the servers in the network and adjusts the behavior of the traffic accordingly. This technique is called a customized flow routing. The Plug-n-Serve application is based on LOBUS (Load Balancing Over Unstructured Networks) algorithm for traffic management. The application reduces the response time of web services in unstructured networks. Smart routing, the main part of the application, considers the server load and path congestion to (re)direct traffic. Figure 1.2 shows the architecture of Plug-n-Serve application. There are three functional units that manage the traffic flow when any web request arrives: flow manager, net manager and host manager. The flow manager is a controller that manages and controls the flow based on the load balancing algorithm used. The net manager collects the network topology and link usage by querying the switches. The host manager keeps track of the load on each server and its state. It also monitors the newly added servers in the network. By using OpenFlow protocol, application collects the data about entire network, including the CPU utilisation and server state. Based on this data, traffic redirection decisions are made by the flow manager. The advantage of Plug-n-Serve is that it reduces the response time for requests from clients by dynamic addition or removal of paths and switches [2].

1.2.2 ElasticTree

The ElasticTree application handles the workload of a data center dynamically. It operates at lower cost and saves the energy by switching off the elements which are not required. It monitors the traffic load continuously and adjusts the active elements like switches, links, ports according to that. Figure 1.3 shows the architecture of

Fig. 1.2 Plug-n-Serve architecture

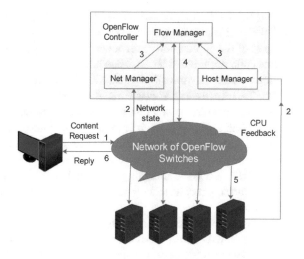

Fig. 1.3 ElasticTree architecture

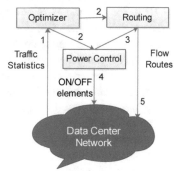

ElasticTree application. It is comprised of optimizer, routing and power control modules. The optimizer module gets the network state and topology information from the network. It then decides which subset of the network can fulfill the current traffic condition and sends this subset to the routing and power control modules for further processing. The routing module then creates the flow routes based on the input data and forwards them to OpenFlow switches. The power control module receives status information from the elements like switches, ports, links to satisfy the current traffic needs and switches off the elements which are not required. Experiments have shown that it can scale the data centers and can save 50% of the energy consumption [3].

1.3   Threat Analysis Methods

Several threat modeling tools and methodologies are used for security analysis, such as PASTA [7], Trike [8], Attack tree [9], UMLSec [10], OCTAVE [11], and Misuse

Cases [12], CC (Common Criteria) [13], DREAD [14], CORAS [15], STRIDE [16]. STRIDE is chosen for the security analysis as this study focuses on the security analysis of the applications from a whole system perspective. We do neither use the source code nor the insider (e.g. designer or architect) view for the analysis. Below are the reasons why the other threat modeling methodologies are not used.

PASTA (Process for Attack Simulation & Threat Analysis) is an attack simulation methodology which is suitable for designers and developers in a organization. To do the threat analysis using this method, the analyst needs to know the definition, technical scope of the application and the implementation details of the system as an insider.

The Trike is a threat modeling tool which is suitable in the design phase, because it is a requirement centric method. The participation of stakeholders is necessary in this method.

Another threat modeling methodology is Attack Tree, which is available as an open source as well as commercial versions. Since it is a attacker centric method rather than a system centric once, it is not a good choice for the entire system analysis.

UMLSec is a model-based approach, where each component of the system is analyzed with various UML stereotypes. However, to be able to use this approach, one must know the source code which is not the aim of this study.

OCTAVE is a risk assessment tool, where an analysis team consisting of experts from various departments is required.

Misuse cases consist of various business process modeling tools where the threat analysis is done based on the expert guidance of various fields like architecture, design and testing which is not possible here while analyzing the system alone.

Common criteria is a framework for the security evaluation of information technology, which is intended for big organizations. It does not specify any standard rules and does not directly provide a list of product security requirements or features.

DREAD is used for risk assessment which is subjective in nature while giving ratings to the threats and moreover the model itself is out of service now.

Risk analysis methods like CORAS need regular customer interaction for the security analysis, which is not feasible in this case.

Chosen Method—STRIDE

STRIDE is a threat modeling tool, proposed by Microsoft in 1999. By using this methodology, a system can be analyzed without considering its implementation, which makes it the most appropriate candidate for this security analysis. STRIDE is an acronym for:

Spoofing: Impersonating a user by illegitimately accessing or using his/her authentication information is spoofing. This is a threat against authentication.

Tampering: Tampering is the data modification for a malicious purpose which is a threat against data integrity.

Repudiation: Repudiation means that users deny performing the action that was done by them which violates the property of non-repudiation.

Table 1.2 DFD components

Type	Component	Description
Process	Circle	Denotes computation or programs run by the computer
Data flows	Arrow	Shows data in motion
Interactors	Rectangle	Represents endpoints of the systems: people, web services, servers etc.
Data store	Two parallel Lines	Represents files, registry, keys, databases etc.
Trust boundary	Dotted Line	Denotes trust boundary between trusted and untrusted elements

Table 1.3 Threat matrix

DFD component	Threat categories					
	S	T	R	I	D	E
Process	X	X	X	X	X	X
Data flows		X		X	X	
Interactors	X		X			
Data store		X		X	X	

Information Disclosure: Data becomes available to the user who is not supposed to have. This is a threat to the confidentiality of the data.

Denial of Service: When legitimate users cannot access the data or a service, then it is called a Denial of Service (DoS). This is a threat against service availability.

Elevation of Privilege: An unauthorized user gets the access rights, which violates the authorization property of a system.

To analyze a system using STRIDE, it is decomposed using a Data Flow Diagram (DFD). A DFD represents the components of a system architecture and its interaction with the internal and external components. These components are briefly described in the Table 1.2. The components are analyzed to check whether they are susceptible to one or more threat categories. STRIDE comes with a threat matrix as shown in Table 1.3, based on which the security analysis of the chosen applications is done in this paper.

1.4 Security Analysis

The SDN routing applications offer dynamic traffic management and control of a data center network. When the entire data center performance and functioning depends on the routing applications, their security becomes an important question. In order to find out the potential threats or vulnerabilities that exist in the application, security analysis is necessary.

In this section, two separate DFDs have been constructed by scrutinizing Plug-n-Serve and ElasticTree architectures respectively. The security of these architectures has been analyzed based on these DFDs.

1.4.1 Plug-n-Serve DFD

In order to evaluate the security of Plug-n-Serve application, it is necessary to construct a DFD which is shown in Fig. 1.4. The components of the DFD are discussed in detail in the following section.

Three functional units of the architecture are located in the controller: flow manager, host manager and net manager. The controller is a logically centralized process which makes all the decisions which affect the behavior of a DC (Data Center) network. So, these units are enclosed into a single process in the DFD.

The Plug-n-Serve architecture has a group of three major components which are represented by the interactors in the DFD: (1) content requesting PCs that send the content request to the network (2) web servers and (3) OpenFlow switches that communicate with the controller.

Furthermore, the communications between the interactors are modeled in the DFD by data flows. The data flows are between content requesting PCs and switches, switches and web servers, controller and the switches, web servers and the controller. The communication between the entities which are located in the controller, assumed to be secured as they reside on the same machine. These entities are flow manager, host manager and net manager.

Finally, the trust boundaries need to be considered among the group of interactors. The controller and switches should neither trust web servers nor the content requesting PCs. There should be a trust boundary between content requesting PCs and OpenFlow switches, switches and web servers, web servers and the controller. The process in the DFD resides in the different network which cannot be trusted by switches and therefore, there is a trust boundary between the controller and switches.

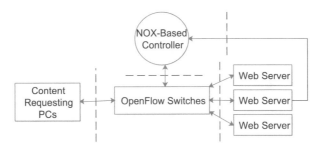

Fig. 1.4 DFD of Plug-n-Serve

All the above observations are considered while deriving the DFD. The security of the components is evaluated based on this DFD.

1.4.2 Plug-n-Serve Analysis

For the security evaluation of the Plug-n-Serve application, the STRIDE tool is used and the following components are evaluated against these threat categories. Table 1.4 represents the threat matrix.

1.4.2.1 Process

Only one process in the DFD is NOX-based controller, which requires protection from all the six STRIDE threat categories.

Spoofing: When an identity is compromised, then an attacker may be able to impersonate as a controller and can take control of the entire routing or load balancing which may redirect the traffic to the malicious servers which are controlled by the attacker. Proper bidirectional authentication or authenticode [17] can be applied as a mitigation technique.

Tampering: The threat of tampering corresponds to the change or replacement of the process binaries. That means, an attacker might change and interpret entire routing process or CPU feedback information. Proper authorization and maintaining an Access Control List (ACL) [18] are suggested to avoid tampering. Digital signature can also be used to ensure authenticity [19].

Repudiation: A malicious user might change the controller program and later denies it. In addition, the user might also deny sending or receiving any data to and from

Table 1.4 Plug-n-Serve threat matrix. Legend: ✓: Denotes threat can be mitigated as architecture provides countermeasure. *: Denotes threat can be mitigated by suggested method

Type	Component	Threat categories					
		S	T	R	I	D	E
Process	NOX-based controller	*	*	*	*	*	*
Data flows	Client PC-switches		*		*	*	
	Controller-switches		*		*	*	
	Switches-web servers		*		*	*	
	Web servers-controller		*		*	*	
Interactors	PC-controller	*		*			
	Controller-switches	*		*			
	Switches-web servers	*		*			

the network. This may result in the improper balancing of the load. As a mitigation technique, the usage of timestamp [18] and maintaining audit trails are recommended.

Information Disclosure: The flow manager collects information about the network topology, CPU utilization and server state. The host manager has the information about the loads of each web server. Information disclosure is about extracting all of these secret data. Encryption of the information is required to avoid this (considering no spoofing or tampering of the data is done). Use of privacy-enhanced protocols like Transport Layer Security (TLS) is able to mitigation from this threat.

Denial of Service: In the context of the process, a successful DoS attack can shut down the process or can deny access to the process for the legitimate users. As a result, OpenFlow switches will not get any routing information from the controller. In such a scenario, controller replication technique can provide data to the switches and can avoid the complete shutdown of the network. Also, packet filtering firewalls or authorization using IP restrictions can be used as a mitigation technique. The flooding attack can be detected using mechanism suggested in [20].

Elevation of Privilege: If an attacker gets privileged access, then the entire routing process can be changed which can stop or destroy the entire system. This can be avoided by running the processes with the least amount of privilege which can assure user rights or resource access permission e.g. use of CPU, memory or network [21].

1.4.2.2   Data Flows

The data flows are prone to tampering, information disclosure and DoS attack. Between content requesting PCs and the network of OpenFlow switches

Tampering and Information Disclosure: Malicious information can be injected in the controller or information may be sniffed from the controller response. Use of TLS would be sufficient to avoid these threats which offers confidentiality by symmetric data encryption and authentication by handshake [2]. Hence, the use of TLS is recommended to avoid these threats.

Denial of Service: The DoS attack can cause unavailability of network of switches if the flooding is done. As a result content requests that are sent by PCs will not be served. Filtering techniques like packet filtering firewalls is one of the solutions. Other techniques like AVANT-GUARD [22], FLOW-GUARD [23] can be explored to avoid flooding.
Between the controller and OpenFlow switches

Tampering: The attacker may modify routing information sent from the controller to the switches while in transit and can force the switches to route the traffic to the malicious server (if there exists any server which is controlled by the attacker). Furthermore, if an attacker modifies the network statistics data sent by switches to the controller then the controller may send improper routing information back to the switches. As TLS option is available in the OpenFlow protocol, this option can be enabled to mitigate the threat. The technique like appropriate authorization or digital signatures can also be used for mitigation.

Information Disclosure: If the information about network statistics and routing feedback sent to the controller is disclosed, then an attacker may get knowledge about the load balancing. An attacker can use this information to build their own application. As discussed above, enabling TLS assures confidentiality by symmetric data encryption and authentication by handshake which can avoid information disclosure.

Denial of Service: As a result of DoS, switches may not get any routing information from the controller. The application directs the data center for routing of requests using this data flow and DoS can make entire system stop functioning. Packet filtering firewalls or ACL is suggested to avoid this. Furthermore, data flow and bandwidth control methods can be used as mitigation techniques.

Between the switches and web servers
The OpenFlow switches assign the content requests to the web servers based on the load on the server. The information flow between them needs to be analyzed.

Tampering: An attacker may change the routing information or may send malicious data in the request. This can result in system crash. Digital signature or use of TLS is suggested to avoid this.

Information Disclosure: Amongst others, this information flow contains data about the server statistics. After observing several such information flows, an attacker can find the pattern of request allocation or less loaded server. This can lead to further attacks on a particular server. TLS option could be configured to mitigate this threat.

Denial of Service: If an attacker floods with the requests to the web server, the web server may crash and will not serve any content request. Filtering techniques or IP restrictions can avoid this threat. Also, data flow and bandwidth control methods are recommended as mitigation techniques.

Between web servers and the controller

Tampering: The web servers send their CPU usage information to the controller, based on which the controller decides how to route the requests. Tampering with this information can result in improper load balancing. The IPSec protocol may be used as a mitigation technique.

Information Disclosure: In order to mitigate this threat, ACL along with encryption techniques could be used.

Denial of Service: If an attacker performs the DoS attack successfully, then the web servers will not be available. Mitigation techniques like filtering or IP restriction is suggested to avoid this.

1.4.2.3 Interactors

The interactors are vulnerable to spoofing and repudiation. The Plug-n-Serve application does not provide any mechanisms that can prevent from spoofing or repudiation.

Content requesting PCs

Spoofing: The threat of spoofing corresponds to the impersonation as a valid user. If an attacker is able to impersonate, he can perform several attacks, e.g. DoS attack. Use of bidirectional authentication or use of Kerberos is suggested as a mitigation technique against this threat.

Repudiation: If an evil person acts as an inside attacker, he can deny sending any malicious requests to the controller. Proper audit trails and timestamps should be maintained at the controller and the web server to avoid this.

Web servers

Spoofing: If an attacker can spoof a web server, then he can send the wrong CPU utilisation information to the controller. This can lead the controller to take wrong decisions based on this wrong data. Also, a web server can send the malicious content e.g. malware which can harm the controller or the content requesting PC. The mitigation technique is to use an appropriate authentication mechanism like Kerberos. Also the use of firewall and deep packet inspection are suggested to mitigate this threat.

Repudiation: A user may modify the functionality of the web server or can change the data and later deny it. An attacker may send malicious data to the controller and later deny it. Mitigation techniques like digital signature, timestamps are suggested to avoid this.

OpenFlow switches

Spoofing: If an attacker can impersonate as an OpenFlow switch, he can change the network statistics that are sent from OpenFlow switches to the controller which can lead controller to take wrong decisions. An attacker can also read and interpret the data exchanged between them. The use of certification and appropriate authentication like IPSec is recommended to mitigate this.

Repudiation: Switches can deny receiving any data from the controller. Proper logging of the data and timestamps are recommended for the mitigation of this threat.

1.4.3 DFD of ElasticTree

As the first step of security analysis, DFD is derived from the ElasticTree architecture. The optimizer, routing and power control are the three modules in the application.

Fig. 1.5 DFD of ElasticTree

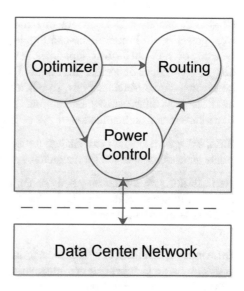

These modules are part of a single system and contained in the same physical box. This is represented as a single process in the DFD.

The data center network communicates with the application to get the routing and switch ON/OFF information. This information flow is modelled in the DFD by the data flow. Additionally, the optimizer, routing process and the power control unit communicate with each other, but as the three processing modules are part of the same system, communication between them are assumed to be trusted.

The trust boundary is placed between an application process and DC network, both reside on the different network. Figure 1.5 shows the DFD derived from all the above observations.

1.4.4   Analysis of ElasticTree

Security analysis of ElasticTree application is done using STRIDE. The threats according to the Table 1.3 are discussed here and mitigation techniques are suggested. Table 1.5 shows the summary of the security analysis.

1.4.4.1   Process

The first component is the ElasticTree application process which is susceptible to all the six threat categories.

Spoofing: As discussed earlier, spoofing in the case of a process can attack the integrity of the process binary. If an attacker impersonates as an optimizer, he may be able to operate a DC network in a malicious way. If an attacker takes control over power control module, he can toggle the power states of the active elements wrongly or even can destroy/shut down the entire network. The application provides an option to reduce or prevent this threat [3] i.e. routing and optimizer module should be separated on different servers, so when the optimizer module is crashed, routing can switch ON all the active elements and application can still work. Other techniques like bidirectional authentication or use of IPSec can mitigate this threat.

Tampering: An attacker can change the application program. The use of an appropriate authorization or a digital signature is recommended to mitigate this threat.

Repudiation: An insider may deny changing the programming of the modules. An attacker might collect the data, including the traffic statistics and later deny receiving it. The use of the timestamp and audit trails are suggested as mitigation techniques for this.

Information Disclosure: If the information about optimizer, routing or power control is disclosed, an attacker can get information about network topology, traffic matrix and active elements. An attacker can use this information for a further attack on the network or can build his own application. Encryption methods should be used to avoid this type of threat (considering that there is no tampering or repudiation of data). Also, proper authorization by ACL and use of TLS are suggested to mitigate this.

Denial of Service: The DoS attack would cause an entire application to stop working. This threat can be mitigated by the deployment design of the application (for example, several controllers). The primary controller may send periodic updates to the other controllers. If a redundant controller doesn't receive the update in time, it takes the charge. As a result, a load balancing technique which is provided by ElasticTree application will not be applied, but the DC network can still work by keeping all the elements active [3].

Elevation of Privilege: This threat can be mitigated by running the processes with the least privilege possible.

Table 1.5 ElasticTree threat matrix. Legend: ✓: Denotes threat can be mitigated as architecture provides countermeasure. *: Denotes threat can be mitigated by suggested method

Type	Component	Threat categories					
		S	T	R	I	D	E
Process	NOX-based controller	*	*	*	*	✓	*
Data flow	PC-controller		*		*	*	
Interactor	Controller-switches	*		*			

1.4.4.2   Data Flow

The data flow is between the application and DC network, which is critical for the functionality of the application.

Tampering: The attacker can change the data in transit between DC network and application. If an attacker is able to perform a change in network topology or traffic matrix, it can force an optimizer to produce wrong results. This can further result into routing the data through malicious servers or can toggle the active elements on/off states wrongly. Use of digital signatures or message authentication is suggested to avoid this threat.

Information Disclosure: The threat here is that the information exchange between controller and application will be disclosed to the attacker. This threat can be mitigated using TLS. The OpenFlow protocol provides an option to configure the TLS, using which a secure connection can be established.

Denial of Service: The data flow needs to be protected against DoS attacks because it corresponds to the flooding of the process or DC network. In this case, DoS attack would prevent an optimizer to get any data about network traffic and a DC network to get any routing information. As a result, the toggle of the power states of active elements will not happen. If the secondary controller takes over the system in the absence of the primary controller, then it can still serve the requests and can run the application. Other forms of DoS attacks could be avoided by using appropriate authentication mechanisms.

1.4.4.3   Interactors

The interactors are prone to spoofing and repudiation. The DC network is the interactor in this DFD which interacts with the application.

Spoofing: Spoofing of a DC network can change entire DC network programming. This can result in providing wrong information to the optimizer or changing of entire functioning of the network. This can be avoided by using appropriate authentication techniques like IPSec's authentication header.

Repudiation: An insider attacker can change the programming of a DC network and can later deny changing it. Usage of audit trails and timestamps are recommended when the controller sends data to the DC network and vice versa.

1.5   Conclusion

The applications based on SDN have greater advantages, but at the same time security of such an application is also an important discussion. In this paper, Plug-n-Serve and

ElasticTree applications are analyzed using STRIDE. Both the applications support data centers to reduce the operating cost by saving energies. In the paper, analysis of these applications with STRIDE is illustrated and mitigation mechanisms are suggested. Similar to the STRIDE, different other threat analysis tools/methods can be used together for further analysis, e.g. a threat can be further analyzed by using an attack tree. Applying security features which are suggested in this paper might be time-consuming or would increase the operating cost, but mitigation of these threats is necessary to avoid the attacks on the system. The analysis performed in this paper shows that the ElasticTree application provides some good mitigation techniques like the use of multiple controllers but in the case of Plug-n-Serve application, security measures like use of timestamps, privacy enhanced protocols are not the part of the design.

Security requirements need to be considered during designing or building stage of any application. The application designers, architects, testers may have different views about the security of these applications, which can be considered to make applications more secure. The intention of this work is not to analyze the security vulnerabilities of the applications through static/dynamic code analysis or pen testing, which are left for the future.

References

1. McKeown Nick et al (2008) OpenFlow: enabling innovation in campus networks. ACM SIG-COMM Comput Commun Rev 38(2):69–74
2. Handigol N et al (2009) Plug-n-Serve: load-balancing web traffic using Open-Flow. ACM Sigcomm Demo 4(5):6
3. Brandon H et al (2010) ElasticTree: saving energy in data center networks. In: NSDI, vol 10, pp 249–264
4. Rolbin M (2013) Early detection of network threats using Software Defined Network (SDN) and virtualization
5. Al-Fares M et al (2010) Hedera: dynamic flow scheduling for data center networks. In: NSDI, vol 10, pp 19–19
6. Kakadia D, Varma V (2012) Energy efficient data center networks—A SDN based approach. IBM Collaborative Academia Research Exchange
7. Real World Threat Modeling Using the PASTA Methodology (2012) OWASP. Technical report. https://www.owasp.org/images/a/aa/AppSecEU2012PASTA.pdf. Accessed 25 Sep 2017
8. Saitta P, Larcom B, Eddington M (2005) Trike v. 1 methodology document [draft]. http://dymaxion.org/trike/Trike_v1_Methodology_Documentdraft.pdf. Accessed 25 Sep 2017
9. Schneier B (1999) Attack trees. Dr. Dobb's J. Technical report https://www.schneier.com/academic/archives/1999/12/attack_trees.html. Accessed 25 Sep 2017
10. Jurjens J (2002) UMLsec: Extending UML for secure systems development. In: International conference on the unified modeling language. Springer, pp 412–425
11. Alberts C et al (2003) Introduction to the OCTAVE approach. Carnegie Mellon University, Pittsburgh, PA
12. Alexander Ian (2003) Misuse cases: use cases with hostile intent. IEEE Softw 20(1):58–66
13. Lohr H et al (2009) Modeling trusted computing support in a protection profile for high assurance security kernels. In: International conference on trusted computing. Springer, pp 45–62
14. Qualitative Risk Analysis with the DREAD Model (2014) Technical report. http://resources.infosecinstitute.com/qualitative-risk-analysis-dread-model. Accessed 25 Sep 2017

15. Lund MS, Solhaug B, Stolen K (2010) Model-driven risk analysis: the CORAS approach. Springer Science & Business Media
16. Threat Modeling with STRIDE (2015, April 16) Technical report https://www.webtrends.com/blog/2015/04/threat-modeling-with-stride/. Accessed 25 Sep 2017
17. Authenticode (2015, April 16) Technical report. https://msdn.microsoft.com/en-us/4.library/ms537359(v=vs.85).aspx/. Accessed 25 May 2017
18. LeBlanc D, Howard M (2002) Writing secure code. Pearson Education
19. Housley R (2009) Digital signatures on internet-draft documents
20. Braga R, Mota E, Passito A (2010) Lightweight DDoS flooding attack detection using NOX/OpenFlow. In: 2010 IEEE 35th conference on local computer networks (LCN). IEEE, pp 408–415
21. Least Privilege (2015, April 16) Technical report. https://www.owasp.org/index.php/Least_privilege/. Accessed 25 Sep 2017
22. Shin S et al (2013) AVANT-GUARD: scalable and vigilant switch flow management in software-defined networks. In: Proceedings of the 2013 ACM SIGSAC conference on computer & communications security. ACM, pp 413–424
23. Hu H et al (2014) FLOWGUARD: building robust firewalls for softwaredefined networks. In: Proceedings of the third workshop on Hot topics in software defined networking. ACM, pp 97–102

Chapter 2
Security Analysis of SDN Cloud Applications

Ankush Chikhale and Rahamatullah Khondoker

Abstract Recently with the emergence of Software Defined Networking (SDN), cloud environments have gone through modifications as traditional data centers adopt SDN as a network management solution. As cloud networking platform provides great power to configure networks in cloud, there is also a downside that intruders and hackers may control the network functionality which may lead to more damage than in legacy networks. Even though cloud networking providers implement the most of the security standards, data storage and important files on external service providers may lead to risk. The ease in procuring and accessing cloud services can also give users the ability to scan, identify and exploit loopholes and vulnerabilities within a system. For instance, in a multi-tenant cloud architecture where multiple users are hosted on the same server, a hacker might try to break into the data of other users hosted and stored on the same server. However, such exploits and loopholes are not likely to surface and the likelihood of a compromise is not great. Understanding traffic flows will extract issues out and methods can be suggested dealing with it. Security concerns here are highly expanded attack that includes the control and data plane. Security challenges are unique to clouds that differ from SDN. In this paper, SDN cloud applications are compared and analysis of three applications such as Meridian, CloudNaaS and HPE Virtual Cloud Network are performed. Main factor for choosing the three applications are their market share and wide deployment. The architecture of these applications are explained and security analysis is done using a threat analysis tool called STRIDE. We suggest some mitigation techniques for the well known threats like spoofing, tampering of data, repudiation and also check if the application has in-built countermeasures against these threats.

A. Chikhale (✉)
Department of Computer Science, TU, Darmstadt, Germany
e-mail: chikhaleankush@gmail.com

R. Khondoker
Fraunhofer SIT, Darmstadt, Germany
e-mail: rahamatullah.khondoker@sit.fraunhofer.de; r.khondoker@yahoo.com

© Springer International Publishing AG 2018   19
R. Khondoker (ed.), *SDN and NFV Security*, Lecture Notes in Networks
and Systems 30, https://doi.org/10.1007/978-3-319-71761-6_2

2.1 Introduction

The cloud should deliver a hosting environment that is immediate, flexible, scalable, secure and available while saving corporations money, time and resources. Five essential cloud characteristics are on demand self-service, broad network access, resource pooling, rapid elasticity and measured service. The cloud itself is a set of hardware, networks, storage, services and come together to deliver computing as a service.

SDN decouples the network control and forwarding functions, enabling the network control to become directly programmable [25] as shown in Fig. 2.1. Although such decoupling is beneficial, SDN security issues such as unauthorized controller access, controller-switch communication flood could be used in cloud environments to harm client applications and network performance. There are lot of cloud computing issues which are already known. These issues are categorized as security standards, network, data, access control and cloud infrastructure [17]. As discussed in [6], from a non-exhaustive survey for cloud issues distribution, author concludes that 18 percent of the overall distribution of cloud issues are related to networking. This motivates the analysis of SDN cloud applications from the security point of view.

Software-Defined Networking (SDN) overcome the above mentioned disadvantages and therefore has become one of the most important networking architectures for the management of networks in cloud computing. SDN gives several benefits over traditional networking such as directly programmable network control, centralised network intelligence in SDN controllers, programmatic configuration, open standards-based and vendor-neutral architecture.

OpenFlow [24] is an open and widespread protocol used as communication interface [33] between SDN control and data planes. OpenFlow allows SDN controllers to access the data plane of network device. This enhanced level of access allows administrators to dynamically change the way traffic flows through the network. The OpenFlow protocol uses a standardized instruction set, which means that any

Fig. 2.1 SDN architecture

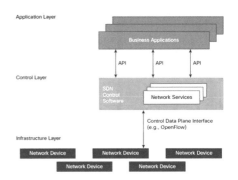

OpenFlow-enabled controller can send a common set of instructions to any Open-Flow enabled switch regardless of vendor.

Server and storage virtualization are important factors in cloud computing and hence SDN security and cloud could be so easily bundled together. SDN falls beyond virtualization aspect of network infrastructure. Understanding traffic flows will extract issues out and various methods can be suggested dealing with it. Security concerns here are highly expanded attack that includes the control and data plane. Security challenges are unique to clouds that differ from SDN and should be discussed.

The rest of the paper is structured as follows: In Sect. 2.2, threat analysis framework is discussed and briefly explained. In Sect. 2.3 Meridian, CloudNaaS and VCN SDN are explained and detailed security analysis of these applications are done in Sect. 2.4. Followed by this, in Sect. 2.5 results of the analysis are discussed and Sect. 2.6 concludes the paper.

2.2 Threat Analysis Frameworks

Threat modeling is a procedure for optimizing network, application and Internet security by identifying objectives, vulnerabilities and then defining counter measures to prevent or mitigate the effects of threats to the system. There are several methods available to analyze security of a system, such as PASTA [26], Trike [7], Attack Tree [28], UMLsec [16], OCTAVE [13], Misuse Cases [1], Common Criteria [20], CORAS [31], DREAD [14] and STRIDE [30].

PASTA is an attack simulation methodology suitable for designers and developers in the enterprises or organizations where user needs to know the definition, technical scope of the application and system from inside to work on threat analysis. *Trike* is a threat modeling tool which is suitable for design phase as it is requirements centric methods and involves stakeholders. *Attack Tree* which is available as open source and commercial software but it is attacker oriented than system oriented. *UMLsec* is a model based approach where each component of a system is analyzed with various stereotypes which requires the source code for the analysis.

OCTAVE is a risk assessment tool for organizations where an analysis team of experts from various departments are required for the analysis which makes it unsuitable for our analysis. *Misuse Cases* is a business process modeling tool based on expert guidance of various fields like architecture, design, and testing. *Common Criteria* is a framework for security evaluation of information technology which is intended for big organizations. It does not specify any standard rules and does not directly provide a list of product security requirements or features for specific products. *DREAD* is also used for risk assessment, but it is subjective in nature when giving ratings to the threats. In risk analysis methods, *CORAS* is used for organizational purpose as it needs customer interaction.

In contrast to the above mentioned tools, by using the *STRIDE* method, it is possible to categorize threats associated with an architecture even without the need of an actual implementation.

STRIDE is an acronym for six categorizes of threats: Spoofing, Tampering, Repudiation, Information Disclosure, Denial of Service, and Elevation of privileges.

- Spoofing: Impersonation as someone else pretending to be something or someone.
- Tampering: Unauthorized change of data—modifying data on disk, on a network or in memory.
- Repudiation: Associated with users who deny performing an action keeping no traces.
- Information Disclosure: Exposure of information to unauthorized persons.
- Denial of Service: Absorbing resources needed to provide service, causing unavailability of services.
- Elevation of privilege: An unprivileged user gains privileged access.

To analyze an application, it should be decomposed into components of a DFD (Data Flow Diagram) and then each component is analyzed against the specific threats for its type. The five types of components of a DFD can be found in Table 2.1. Each of the described elements of a DFD is only susceptible to a subset of the threat categories of STRIDE [5]. For example, interactors are only prone to spoofing and repudiation threats. The mapping of security property to STRIDE threats and threat matrix can be found in Table 2.2. As soon as the threats to the application are identified, a proper mitigation methods can be defined. The mitigation methods proposed in this paper are only suggestions for Meridian and CloudNaaS and some of them are already used by HP VCN.

Table 2.1 Components of a DFD

Item	Symbol
Process	Circle
Data flow	Arrow
Data store	Two parallel horizontal line
Interactors	Rectangle
Trust boundary	Dotted line

Table 2.2 Threat matrix taken from [5] X denotes threat category for particular element

Security property	Threat category	Interactors	Processes	Data store	Data flows
Authentication	Spoofing	X	X		
Integrity	Tampering		X	X	X
Non-repudiation	Repudiation	X	X		
Confidentiality	Information disclosure		X	X	X
Availability	Denial of Service		X	X	X
Authorization	Elevation of privilege		X		

Table 2.3 Overview of SDN Cloud application

SDN Cloud application	Functionality	Developed by in year	Deployed as	Proprietary or Opensource
HPE VCN SDN [10]	Deployment of dynamic policy of the network	HPE, 2014	Saas	Proprietary
EOS [13]	Extensible,event driven operating system	Arista, 2011	Saas	Proprietary
Openstack [24]	OS providing compute, storage and resources	OpenStack community, 2010	Saas	OpenSource
CloudNaaS [26]	Extensible networking platform	IBM, 2011	Naas	NA
Microsoft Azure [27]	Provides applications and infrastructure	Microsoft, 2009	Iaas, Paas	Proprietary
Zimory Cloud suit [28]	Provides distributed, scalable, decoupled infrastructure	Zimory, 2008	Iaas	Proprietary
Vmware vCloud suit [29]	Provides infrastructure and its management	Vmware, 2013	Iaas	Proprietary

2.3   SDN Cloud Applications

In the past couple of years, several SDN cloud applications have been developed. A summary of the applications can be found in Table 2.3. Each application is described by its name, functionality, developed by (company/institute) /Year, deployed as and proprietary/open source. All the applications in Table 2.3 are based on the OpenFlow protocol.

Two SDN cloud networking platforms Meridian and CloudNaaS and one cloud application HP VCN SDN are chosen for security analysis in this paper. Main factors for choosing these three applications are their market share, and wide deployment. Meridian and CloudNaaS are chosen because they provide service-level cloud networking model and are deployed widely using network as a service to end users. Both are developed by IBM T. J. Watson Research Center in 2011 and 2013 respectively. Even though these applications are widely deployed but not much thorough research and analysis has been done on these applications. In order to provide sufficient context for the security analysis, concepts and architecture of all three applications are briefly described, following description of data flows and processes with the help of data flow diagrams. On the basis of this information, CloudNaaS, Meridian and HP VCN SDN applications will be analyzed using the STRIDE framework.

2.3.1 *Meridian*

Meridian is an extensible multi-threaded SDN platform for cloud networking application. It supports service-level model and provides multiple options for configuring
virtual networks on the underlying physical network. Meridian consists of three
main logical layers: Application Programming Interfaces (APIs), network orchestration and underlying network devices. Network applications are at the top of the
stack, as consumers of the APIs.

The Meridian architecture diagram is shown in Fig. 2.2. The topmost layer of the
architecture consists of the abstract API to be able to interact with the network. The
API provides access to the higher layer cloud applications to request policy based
connectivity between logical groups of virtual servers i.e., details such as construction
of virtual network with virtual machines, along with policies for controlling an access,
prioritizing traffic or traversing middle-boxes.

Network is represented in the form of graphs where the network elements are
represented as graph nodes and a relationship between elements are represented by
a graph edge. Network orchestration layer performs a logical-to-physical translation
of commands issued through the abstraction layer above and convert these API calls
into the appropriate series of commands on the underlying network. The lowest
layer consists of logical driver that interfaces to OpenFlow devices to create virtual
networks and accompanying services. Meridian accepts configuration rules from

Fig. 2.2 Meridian
Architecture(modified and
redrawn)

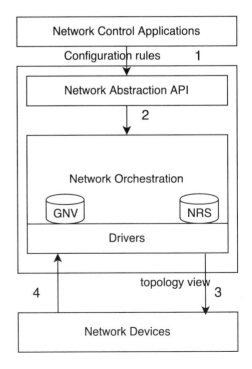

network control applications for underlying network devices. Meridian provides topological views of the dynamic set of underlying network resources to upper layer with the employment of libvert [19].

Libvert is the virtualization API and provides consistent interface for querying and controlling underlying network devices. Meridian also provides as a network abstraction model service, an entity to describe the services on a connectivity path. Users can define a customized routing policy, filter or middlebox traversal policies in a service entity. The communication between different tiers often has different service requirements, for example, restricted communication between tiers using firewalls. Network orchestration layer contains two data stores namely Global Network View (GNV) and Network Runtime State (NRS) which store underlying network view and underlying network state details respectively. Meridian is built on open source Floodlight [8] controller platform. Floodlight is a modular java based OpenFlow controller. Further details about Meridian can be found in [3].

2.3.2 CloudNaaS

CloudNaaS, is an SDN based cloud networking platform for enterprise applications, with the purpose of providing networking primitives for cloud applications. Cloud-NaaS uses NOX controller and is based on OpenFlow protocol. With CloudNaaS, customers are able to deploy their applications on the cloud to access virtual network functions such as network isolation, custom addressing, and service differentiation. CloudNaaS also provides an ability to deploy middle box appliances to provide intrusion detection, caching or application acceleration.

The CloudNaaS architecture consists of two main components namely, the cloud controller and the network controller. The cloud controller manages both the virtual resources and the physical hosts and supports APIs for setting network policies. The network controller is responsible for monitoring and managing the configuration of network devices as well as for deciding placement of virtual machines within the cloud. Network controller is a new component introduced by CloudNaaS in Cloud Management Service (CMS) as compared to other typical clouds. CloudNaaS network controller is implemented on top of the NOX controller using C++.

Cloud controller accepts network policy specifications and virtual machine requests from users. It converts user requirements into a communication matrix. Network controller compiles matrix entries into network level rules. It installs rules on virtual and physical switches through SDN control channels and configures paths. The Network Provisioner (NP) inside a network controller collects information about communication matrix from the cloud controller. The Cloud Monitor (CM) period-ically polls status of switches and links, as shown in Fig. 2.3. Network provisioner and cloud monitor act as data stores in network controller. Network controller also supports network aware Virtual Machine (VM) placement, Quality of Service (QoS) support, real time network monitoring, flexible diagnostics, management and security functions. In network controller, details of placement optimizer and state optimizer

Fig. 2.3 CloudNaaS
Architecture(modified and
redrawn)

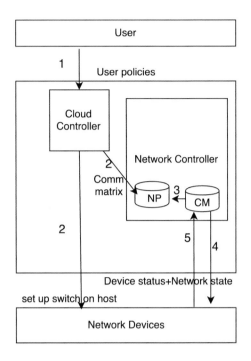

and their interactions with other components are omitted for the ease of evaluation purpose. CloudNaaS also employs pre-computation and caching of alternate paths to reduce the impact of a device or a link failure in the underlying network. Further details on CloudNaaS can be found in [4].

2.3.3 HPE VCN SDN

It is the enhanced networking module of HPE Helion OpenStack. HPE VCN enables the enterprises to securely connect to the cloud and applies its own identity to its cloud environment as shown in Fig. 2.4. Since the Virtual Cloud Network solution is already integrated with OpenStack, public cloud providers can deliver an automated self-service solution to their tenants and enterprises can securely connect their private estate to public cloud environments.

1. **HPE Helion OpenStack**, builts upon OpenStack [4], is an open and extensible scale-out cloud platform that make it easier to build, manage and consume workloads in a hybrid IT environment. It is open by providing the ability to move, integrate and deliver applications across public, private and traditional IT environments. It is secure by offering secure solutions to all Hybrid IT Clouds with the use of Cryptography. It is agile by enabling speedy deployment of either private or public cloud within less span of time [29].

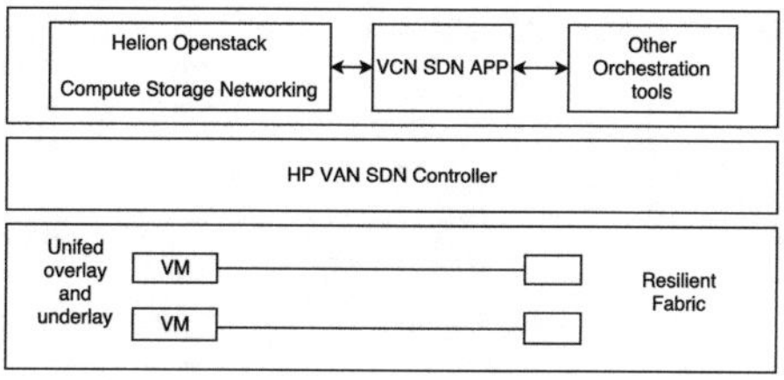

Fig. 2.4 HP VCN SDN
Architecture

2. **HPE VAN SDN Controller** HPEs Virtual Application Networks (VAN) SDN Controller is a building block of HPEs virtualized data center solution. The HPE SDN Controller manages policy and forwarding decisions which are communicated to the OpenFlow-enabled switches in the data center [12].
3. **Network devices** The HPE FlexFabric 7900 Switch Series is the next generation compact modular data center core switch designed to support virtualized data centers and evolutionary needs of private and public clouds deployments.

2.4 Security Evaluation of the Cloud Applications

2.4.1 Security Evaluation of Meridian

To evaluate Meridian with STRIDE, a DFD is constructed as shown in Fig. 2.5. The following assumption is made for the evaluation purpose: The three layers of Meridian are contained in the same physical box and assumed as a single Meridian process. STRIDE can be applied on Meridian process as a single process and thus it is not necessary to evaluate data flow between the Meridian process layers. The network control application, Meridian architecture, and network devices are distributed in cloud and the data flow between them cannot be trusted, hence there is a trust boundary between the network control application and the Meridian process and between the Meridian process and the network devices. The two data stores global network view (GNV) and network run time state (NRS) in network orchestration layer are assumed on the same physical device inside the meridian architecture. The two actors in the DFD are the network control applications who interact with API provided by cloud controller and network devices who enable interaction via API provided by drivers.

For the evaluation using the STRIDE methodology, the following components and corresponding threats and mitigation methods are discussed below. Table 2.2 shows a summary of the security evaluation of Meridian.

Meridian architecture as a single process:
Spoofing: A malicious user may impersonate an authenticated network control application and can make changes to Meridian process binary. Only admin should be able

Fig. 2.5 Meridian DFD

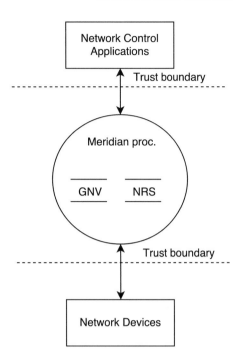

to change the process binary. This can be mitigated by applying proper authentication mechanism such as Authenticode [2] to validate code that has been signed by the admin user.

Tampering with data: A malicious user an impersonate an authenticated network control application and can modify, replace and view Meridian process binary. For mitigation, access control list (ACL) [9] to specify the access rights could be preferred over digital signature and message authentication codes, resulting in less messaging overhead each time Meridian process is run.

Repudiation: A user can change the binary of Meridian without keeping any proof, which may be difficult to track later in time. User activity can be tracked by time stamps, audit trails. Digital signatures [9] are not preferred as they may lead to overhead for each action in Meridian process.

Information Disclosure: An attacker may extract Meridian process binary and view the Meridian process working details. Encryption techniques such as block ciphers can be used as a mitigation plan. Block cipher [9] processes a block of input data and produces a cipher text block of the same size resulting in less memory overhead.

Denial of Service (DoS): A malicious user may consume network resources by excessive use of the underlying network. For example, a user can write overlapping or unnecessary network configuration rules thereby occupying all the network switches/bandwidth resulting in shutting down of Meridian process or denying access to underlying resources. For mitigation in such cases, disk quotas [9] can be used to prevent excess disk consumption. Suitable authentication mechanism such as digest authentication can be applied for administrative entities for shutting down Meridian process or locking resources.

Elevation of Privileges: A user can access/alter/destroy Meridian process binary. Input validation [9] for the administrators or users running with least privileges could also be used as a mitigation plan.

Data flow between the Network Controller Application to the Meridian Process:
Tampering with data: Data sent to the Meridian process from an authorized network controller application can be changed by an attacker when in transit. Digital signatures can be used as a mitigation plan to check the integrity of data.

Information Disclosure: Network configuration commands while in transit can be read by an attacker, thereby gaining access to underlying virtual network topology requested by an user. TLS [9] can be used to mitigate the risk which will protect network configuration commands confidentiality by using symmetric data encryption and TLS handshake for data authentication.

Denial of Service (DoS): Data flow is represented by network traffic and an attacker can use filtering or throttling to control and modify the network traffic. For example, an attacker can send packets with network configuration details with high frequency to Meridian resulting in bandwidth bottleneck. For the mitigation, bandwidth and data flow control can be applied to limit packet flows.

Data flow between Meridian Process and Network Devices:
Tampering with data: Meridian process discovers underlying topology view by monitoring advertisements sent by the libvert virtualization daemon. Libvert provides remote management using TLS encryption and x.509 certificates [15] and thus threats could be mitigated.

Information Disclosure: Underlying network topology is discovered using libvert. Information disclosure could be mitigated since libvert provides Kerberos [32] and SASL [19] for data authentication.

Denial of Service (DoS): An attacker may not able to send network topology packets on behalf of the Meridian process to network devices for the configuration of the underlying network as Kerberos [19] and SASL provided by libvert will ensure secure authenticated communication between the Meridian process and underlying network devices.

Interactors Network Control Application and Network Devices:
Spoofing: An attacker may act as the Meridian process and may receive policy based connectivity between logical groups of virtual servers on behalf of the authenticated Meridian provider. The user will assume the underlying network is configured according to specification and can expect desired performance, underlying Meridian process being unaware of it. Proper authentication mechanism such as Kerberos can be used to authenticate the identity of Meridian process as well as users. An attacker may not be able to send requests to get the topological view from the network devices on behalf of the Meridian service and gain view of underlying topology since Kerberos and SASL [19] provided by libvert will ensure safe authenticated communication between the Meridian process and underlying network devices.

Repudiation: As network control applications are the initiators of request, the Meridian process might not request any changes to network control applications. TLS as incorporated in libvert [19] will ensure non-repudiation for the communication between the Meridian and the network devices.

Data Store Global Network View (GNV) and Network Runtime State (NRS):

Tampering with data: Data stores global network view (GNV) and network runtime state (NRS) might be altered or deleted by an attacker. ACL can be used to specify access rights for each user or role based access control (RBAC) can be used for mitigation.

Information Disclosure: Data stored in global network view and network runtime state could be read by an attacker, thereby gaining access to underlying virtual network topology. Data encryption can be used as a mitigation plan for data stored in databases.

Denial of Service (DoS): Data stored in global network view and network runtime state could be modified/deleted by an attacker, resulting in DoS to authorized users. Message authentication codes might be used over digital signatures (inefficient with message overhead) for filtering the authorized users (Table 2.4).

2.4.2   Security Evaluation of CloudNaaS

To evaluate CloudNaaS using the STRIDE method, a DFD is built accordingly as shown in Fig. 2.6. For the evaluation purpose, the following assumption is made: the two layers of CloudNaaS are contained in the same physical box. Therefore, it is not necessary to evaluate the data flow inside the CloudNaaS process. STRIDE can be applied on CloudNaaS process as a single process. The two actors in the DFD are the user who interact with API provided by cloud controller and network

Table 2.4 Security analysis of meridian components, * denotes threat can be mitigated by suggested methods, ✓ denotes threat can be mitigated as architecture provides countermeasures to mitigate, - denotes out of scope

Type	Component	Threat					
		S	T	R	I	D	E
Process	Meridian	*	*	*	*	*	*
Data flows	Network Control applications and Meridian		*		*	*	
Data flows	Meridian and Network Devices		✓		✓	✓	
Interactors	Network & Control Applications	*		-			
Interactors	Network Devices	✓		✓			
Data Store	Global Network View		*		*	*	
Data Store	Network runtime systems		*		✓	*	

Fig. 2.6 CloudNaaS DFD

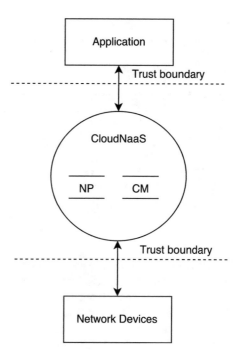

devices who enable interaction via API provided by the network controller. The user, CloudNaaS architecture, and network devices are distributed in cloud and data flows between them cannot be trusted, therefore, there is trust boundary between the user and the CloudNaaS process and between the CloudNaaS process and the network devices, therefore, data flow must be checked for security threats across trust boundary. Network Provisioner (NP) and Cloud Monitor (CM) act as data stores in CloudNaaS architecture. The evaluation of CloudNaaS is performed below in the same way as for Meridian.

For the evaluation using the STRIDE methodology, the following components and corresponding threats and mitigation methods are discussed. Table 2.5 shows the summary of the security evaluation of CloudNaaS.

CloudNaaS architecture as a single process:
Spoofing: An attacker may make changes to CloudNaaS process binary. As mentioned before, only the administrator should be able to change the process binary. This can be mitigated by applying proper security mechanism ensuring the integrity of binary such as message authentication codes.
Tampering with data: A malicious user may impersonate an authenticated user and may modify, replace and view CloudNaaS process binary. As mentioned before, access control list (ACL) can be used for mitigation.
Repudiation: A user might change the binary of CloudNaaS without keeping any proof, which may be difficult to track later in time. User activity could be tracked by time stamps, and audit trails.

Table 2.5 Summary of the security analysis for the CloudNaaS, * denotes threat can be mitigated by suggested methods, ✓ denotes threat can be mitigated as architecture provided countermeasures, - denotes out of scope

Type	Component	Threat					
		S	T	R	I	D	E
Process	CloudNaaS	*	*	*	*	*	✓
Data flows	User and CloudNaaS		*		✓	*	
Data flows	CloudNaaS and Network Devices		*		*	*	
Interactors	User	✓		-			
Interactors	Network devices	✓		✓			
Data store	Cloud Monitor (CM)		*		*	*	
Data store	Network runtime systems		*		*	✓	

Information Disclosure: An user might extract CloudNaaS process binary and secret data can be extracted from it. Encryption techniques can be used to encrypt the binary and prohibiting access by unauthorized entities.

Denial of Service (DoS): A DoS attack may result in shutting down of CloudNaaS process or denying access to underlying resources. For mitigation in such cases, suitable authentication mechanisms can be applied for administrative entities for shutting down CloudNaaS process or locking resources.

Elevation of Privileges: A user with restricted privileges can access/alter/destroy CloudNaaS process binary. This can be mitigated by running process with minimum privileges.

Data flow between the User to the CloudNaaS Process:

Tampering with data: User policies sent to the cloud controller could be modified by an attacker when they are in transit. Digital signatures and message authentication codes can be used as a mitigation plan.

Information Disclosure: An attacker can trap and expose user policies thereby gaining access to underlying network topology view or user specifications for requested topology. TLS incorporated in OpenFlow can be used as mitigation plan.

Denial of Service (DoS): An attacker may send user policy packets with high frequencies to the cloud controller, resulting in bandwidth bottleneck. This may lead to DoS to other users. For the mitigation, bandwidth and data flow control methods could be used.

Data flow between CloudNaaS Process and Network Devices:

Tampering: An insider attacker may change the device state while in transit from network devices to network controller. Network state can be tracked and viewed, while sent from network controller to network devices. Cloud controller messages to setup virtual switch on host can also be modified in transit. IPSec's Authentication Header (AH) could be used as a mitigation plan.

Information Disclosure: An unauthorized user might extract the details of the underlying network topology by sniffing the data packets sent to the cloud monitor.

Encryption mechanisms could be used as mitigation plan.

Denial of Service (DoS): The user can occupy all the underlying network resources by flooding the networks with corrupted packets, resulting in DoS to other users. For mitigation, data flow and bandwidth control mechanisms could be applied.

Interactors—User and Network Devices:

Spoofing: A malicious attacker can act as a CloudNaaS process and may receive network policy specifications and virtual machine requests on behalf of the authenticated CloudNaaS provider. The user will assume the underlying network is configured according to the specification and may expect desired performance, underlying CloudNaaS being unaware of it. Proper authentication mechanism such as Kerberos could be used to authenticate the identity of CloudNaaS process as well as users. An attacker may hijack the network devices and may change the topology view sent by network devices to the network controller. Authentication mechanisms such as Kerberos, IPSec can be used as mitigation plan.

Repudiation: Users are the initiators of request and CloudNaaS is the receiver. Device status information sent from the network devices to the network controller can be denied later in time if no logs are present to verify. A proper logging mechanism could be implemented to mitigate the threat.

Data Stores—Cloud Monitor (CM) and Network Provisioner (NP):

Tampering: Data stored in the network provisioner (communication matrix) and cloud monitor (switches and link status) could be modified/deleted by an attacker, resulting in malfunctioning of the underlying network. ACL can be used to provide access to authorized user.

Information Disclosure: Data stored in the network provisioner and the cloud monitor might be accessed by an attacker, thereby gaining access to underlying virtual network topology. Data encryption mechanisms could be used to mitigate the risk.

Denial of Service (DoS): Data stored in network provisioner and cloud monitor may be modified/deleted by an attacker, occupying all the underlying resources or bandwidth. An authorized user may experience DoS due to insufficient resources and limited bandwidth. Digest authentication and packet filtering firewalls can be used to check authenticity of user policy packets and verification of communication matrix.

2.4.3 Security Evaluation of HPE VCN SDN

To evaluate HPE VCN SDN application with STRIDE, it is first necessary to build the Data Flow Diagram. The DFD of the application is shown in Fig. 2.7. It is drawn with the help of the VCN SDN application architecture diagram which is shown in Fig. 2.4. Being the key element, the VCN SDN application comes as a part of HPE Helion Openstack distribution. We consider here the Helion OpenStack and the VCN SDN app as one process and VAN controller as another. The two actors in the DFD are network devices which send and receive the data and the administrator

Fig. 2.7 VCN application
DFD

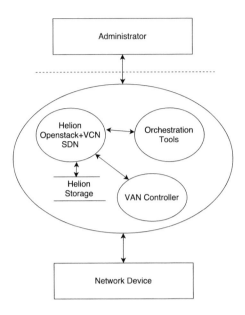

who uses VCN application. Communication flow between administrator and Helion, Helion and VAN Controller, VAN controller and network device should be taken into consideration for the security analysis. There is a trust boundary between the hardware and the VAN controller on one side and between VCN application and the administrator on the other side.

For evaluation using the STRIDE methodology, the following components and corresponding threats and mitigation methods are discussed below. Table 2.2 shows a summary of security evaluation of the HPE VCN SDN application.

Process
At first, the HPE Helion OpenStack with VCN SDN process is taken into consideration. All HPE Helion OpenStack version 2.1 example cloud models ship with Transport layer Security (TLS) [5] enabled on the public API endpoints. TLS protocol can be enabled on initial deployment which mitigates tampering, and information disclosure threats. Encryption mechanisms are also used for passwords and sensitive data, e.g. for secret keys generated by configuration manager, it uses preCrypto libraries and Ansible vaults and openSSL for user supplied passwords. HPE refines access control with AppArmor which is a mandatory access control (MAC) system. Integration with HPE ArcSight provides the ability to monitor, analyze and correlate OpenStack logs to ArcSight logger and supports continuous security monitoring [27].

VAN Controller process—HPE recommends for Fortinet FortiGate firewall for SDN controller security.

Data Flows All incoming traffic (whether it is from the administrator or from the VAN controller) to OpenStack services on the public endpoints are secured using

Table 2.6 Summary of the security analysis for the components of hp vcn sdn, * denotes threat can be mitigated by suggested methods, ✓ denotes threat can be mitigated as architecture provides countermeasures to mitigate, - denotes out of scope

Type	Component	Threat					
		S	T	R	I	D	E
Process	VCN SDN(V)	✓	✓	✓	✓	✓	*
Data flows	(O) ↔ (V)		✓		✓	✓	
Data flows	(V)↔(N)		✓		✓	✓	
Interactors	Administrator(O)	*		✓			
Interactors	Network Devices(N)	*		*			
Data Store	Openstack db		✓		✓	✓	

TLS connections. Use of mandatory access control system (AppArmor) mitigates information disclosure and DoS threats.

Data Stores HPE helion openstack has its own database. As a part of openstack, use of TLS and AppArmor ensure integrity, confidentiality and availability.

Interactors

Network devices—HPE network devices also provide security. HPE uses access control list for filtering traffic to prevent unauthorized access. With authentication and encryption, use of secure shell (SSHv2) protects against IP spoofing and plain text password interception. If network devices are not from HPE, they should also use appropriate authentication mechanism with encryption and digital signatures.

Administrator—To prevent spoofing and repudiation attacks, TLS protocol with MAC system and encryption methods are implemented in the system. Integration with HPE ArcSight reduces the time taken to respond to security breaches, if any and provides faster analysis of logs and events (Table 2.6).

2.5 Discussion and Related Work

Some broadly perceived advantages of using SDN for cloud networking come from the possibility to implement IDS, IPS, firewall, load balancers and much more network functions as software modules in the SDN controller. By doing so, emerging threats can be addressed quickly by the programming of additional software modules.

Meridian and CloudNaaS both provide interposition of middlebox in cloud network such as Deep Packet Inspection (DPI). Although it is worth noting that interposition of middlebox is prone to security threats. Meridian uses Libvert virtualization API and provides consistent interface for querying and controlling underlying network devices. With Meridians network abstraction model service, users can describe services such as filter or middlebox traversal policy on a connectivity path. From the above analysis, Meridian as a cloud controller platform can be proposed over

CloudNaaS, as security issues are well addressed in Meridian as compared to Cloud-NaaS by incorporating Libvert virtualization API and network abstraction model service.

The STRIDE method has been used to accomplish the security analysis of other SDN protocols including OpenFlow, OF-Config, and OSVDB [5], SDN architectures such as PCE, 4D and SANE [18], SDN applications for monitoring and measurement such as sFlow and BigTap [22], SDN cellular applications including OpenRadio and SoftRAN [21], and SDN security applications [23]. All these above related work also provide security perspective for SDN concepts and well tested mitigation techniques to tackle them.

Other than in-built security functions, HP VCN application provides several security products which can be used in combination with existing applications to enhance security. HPE has a portfolio of security products that can be applied to cloud system and the underlying infrastructure components to enhance security of cloud deployments and service offering lifecycle [18]. These comprehensive security solutions are Fortify Software Security Center (SSC), Software Code Analyzer (SCA), WebInspect [10] for Application Security, TippingPoint Next-Generation Intrusion Prevention System (NGIPS), Next-Generation Firewall (NGFW) [11] for Network Security, ArcSight Logger [9] for Security Information and Event Management (SIEM) and HPE Atalla Cloud Encryption [19] for Data / Information Security.

2.6   Conclusion

Two SDN cloud networking platforms, Meridian and CloudNaaS, and one cloud application, HPE VCN SDN, are analyzed using the STRIDE method. This paper analyzed threats to these applications and suggests security mechanisms to mitigate the identified threats. Basic concept of each architecture is explained in brief. Data flow diagrams (DFD) are used to analyze the system from the security perspective. Each security perspective is analyzed in depth and security mechanisms for mitigation are suggested.

New protection mechanisms must be analyzed in depth before their deployment, along with other threat modeling framework such as DREAD, Trike and many others. The security patterns observed can also be applied to other SDN applications, after analyzing in terms of security strength before deployment.

Being in the top organization ranking for cloud providers, HPE has already taken care of many security related issues. HPE Helion OpenStack has many built-in security controls, but the customer must take responsibility for configuring the network devices that integrate Helion services into an existing data center environment. This includes defining firewall rules at the edge of the HPE Helion OpenStack deployment (to protect against external abuse) as well as defining router rules within the HPE Helion OpenStack deployment (to protect against insider abuse or administrative errors).

HP VCN SDN application has taken security measures into consideration but still customer needs not only to configure network devices, operating system and controller appropriately but also should use other security applications to enhance the security. Here, deployment configuration plays an important role. The burden on the customer can be reduced by providing and enabling some of the security features (e.g. TLS) by default.

References

1. Alexander I (2013) Misuse cases: use cases with hostile intent. IEEE Softw 20(1):58–66
2. Authenticode. https://msdn.microsoft.com/en-us/library/ms537359(28v=vs.85)29.aspx (visited on 05/06/2016)
3. Banikazemi M et al (2013) Meridian: an SDN platform for cloud network services. https://doi.org/10.1109/MCOM.2013.6461196
4. Benso T et al (2011) CLOUDNaaS: a cloud networking platform for enterprise applications. In: Proceedings of ACM symposium cloud computing 322(8): 6–24. https://doi.org/10.1145/2038916.2038924
5. Brandt M et al (2014) Security analysis of software defined networking protocols openflow. In: OF-Config and OVSDB in IEEE ICCE 2014 (July 2014)
6. de Jesus WP. Analysis of SDN contributions for Cloud Computing Security. In: IEEE ACM 7th international conference on utility and cloud computing. London, UK, 8–11 Dec 2014
7. Eddington M, Saitta P, Larcom B (2016) Trike v. 1 methodology document [draft]. https://dymaxion.org/ (visited on 06/11/2016)
8. Floodlight. http://www.projectfloodlight.org/floodlight (visited on 05/06/2016)
9. Howard M, LeBlanc D (2015) Writing secure code. Microsoft Press. ISBN: 0-7356-1722-8
10. HP ArcSight Logger. http://www8.hp.com/h20195/v2/GetPDF.aspx/c04447843.pdf (visited on 07/14/2016)
11. HP Atalla Cloud Encryption. http://www.hp.vpm/hpinfo/newsroom/Accessed (visited on 06/19/2016)
12. HP Bridging the data center of today and tomorrow with SDN. http://h17007.www1.hp.com/docs/networking/datacenter/4AA5-1865ENWDiscover-FAQ.PDF (visited on 06/21/2016)
13. HP Fortify Software Security Center v4.30 and HP WebInspect v10.40 Products. https://community.hpe.com/hpeb/attachments/hpeb/ (Visited on 06/21/2016)
14. Improve your network security in 30 days with HPE TippingPoint. https://ssl.www8.hp.com/emeaafrica/en/ssl/(visited on 07/21/2016)
15. Internet X.509 public key infrastructure Time-Stamp Protocol (TSP). https://www.ietf.org/rfc/rfc3161.txt (visited on 07/22/2016)
16. Jürjens J, Hussmann H, Cook S (2012) UMLsec: extending UML for secure systems development. Springer, Berlin Heidelberg. ISBN: 2460- 12-425-200
17. Khalil IM, Khreishah A, Azeem M (2014) Cloud computing security: a survey. In: Computerss 3(1): 1–35. www.mdpi.com/2073-431X/3/1/1/pdf
18. Klingel D et al (2014) Security analysis of software defined networking architectures PCEw. In: Chikhale A, Khondoker R (eds) Asian conference on internet engineering (Nov 2014). https://doi.org/10.1145/2684793.2684796.38
19. Libvert. Libvert: the virtualization API. https://libvirt.org/ (visited on 07/16/2016)
20. Lohr H et al (2009) Modeling trusted computing support in a protection profile for high assurance security kernels. In: International conference on the technical and social economic aspects of trusted computing. Oxford, UK,68 (Apr 2009)
21. Magin D, Khondoker R, Bayarou K (2015) Security analysis of OpenRadio and SoftRAN with STRIDE framework. In: The 24th international conference on computer communications and applications (ICCCN 2015). IEEE, Las Vegas, Nevada, USA (3–6 Aug 2015)

22. Marx R, Dauer P, Khondoker R, Bayarou K (2015) Security analysis of software defined networking for monitoring and measurement sFlow and Big- Tap. In: 10th International Conference on Future Internet Technologies (CFI) (June 2015)
23. Marx R, Tasch M, Khondoker R, Bayarou K (2014) Security analysis of security applications for software defined networks. In: 10th AINTEC 2014. Chiang Mai, Thailand (26–28 Nov 2014)
24. Mckeown N et al (2008) OpenFlow: enabling innovation in campus networks. http://ccr.sigcomm.org/online/files/p69-v38n2nmckeown.pdf
25. ONF. SDN definition. https://www.opennetworking.org/sdn-resources/sdn-definition (visited on 07/15/2016)
26. Real World threat modeling using the PASTA methodology. https://www.owasp.org/images/a/aa/AppSecEU2012_PASTA.pdf (visited on 06/16/2016)
27. Reducing network complexity, boosting performance with HP IRF technology. http://h17007.www1.hp.com/docs/reports/irf.pdf (visited on 07/12/2016)
28. Schneier B (2016) Attack trees. https://www.schneier.com/academic/archives/1999/12/attack_trees.html (visited on 06/11/2016)
29. SDN. Realizing the power of SDN with HP Virtual Application networks. http://h17007.www1.hp.com/docs/interopny/4aa4-3871enw.pdf (visited on 05/14/2016)
30. STRIDE. The STRIDE Threat model. https://msdn.microsoft.com/en-us/ens/library/ee823878(v=cs.20).aspx (visited on 07/15/2016)
31. The CORAS approach to model- driven risk analysis. https://securitylab.disi.unitn.it/lib/exe/ (visited on 06/18/2016)
32. The Kerberos Network Authentication Service (V5). https://www.ietf.org/rfc/rfc4120.txt (visited on 06/17/2016)
33. What is OpenFlow? Definition and how it relates to SDN. https://www.sdxcentral.com/sdn/definitions/what-is-openflow/

Chapter 3
Security Analysis of SDN Applications for Big Data

Parvez Ahmad, Sven Jacob and Rahamatullah Khondoker

Abstract Big Data is a term that describes structured and unstructured large data sets. One of the frameworks to store, process and analyze this data is Apache Hadoop. Software Defined Networking (SDN) enhances the performance aspects of Hadoop by optimizing bandwidth utilization and improving network management. Security attacks on the SDN controller and switches can compromise the whole Hadoop system, that may cause loss or manipulation of valuable data. We selected the three most advanced approaches that focus on accelerating the data transfer between the cluster nodes. FlowComb, Pythia and Hadoop-Acceleration (Hadoop-A) focus mainly on optimizing performance but do not consider any security aspect in their design. This motivates us to analyze the security aspects of these SDN applications. This paper focuses on the analysis of security features with STRIDE threat modeling technique. All approaches need improvements to gain security. We find that Pythia is natively the most secure approach while other approaches can be secured by deploying add-on security mechanisms.

3.1 Introduction

SDN [13] is an approach aimed at making the network agile by separating the conventional data plane from the control plane which allows network engineers and administrators to manage data traffic from a centralized control point. In simple terms, the fundamental idea behind SDN is that multiple ordinary switches are connected to an intelligent controller. Although this separation provides flexibility to the networking world, which enable a whole new level of security threats.

P. Ahmad (✉) · S. Jacob
Department of Computer Science, TU Darmstadt, Darmstadt, Germany
e-mail: in.parvez@gmail.com

S. Jacob
e-mail: crypto@svenjacob.org

R. Khondoker
Fraunhofer SIT, Darmstadt, Germany
e-mail: rahamatullah.khondoker@sit.fraunhofer.de; r.khondoker@yahoo.com

© Springer International Publishing AG 2018
R. Khondoker (ed.), *SDN and NFV Security*, Lecture Notes in Networks
and Systems 30, https://doi.org/10.1007/978-3-319-71761-6_3

The OpenFlow [15] protocol is an open standard protocol developed by Open Networking Foundation (ONF) and is designed for SDN. ONF considers OpenFlow as the first standard communication protocol that enables an SDN Controller to interact with the forwarding plane of network devices. It provides a platform to directly access and manipulate the forwarding plane of network devices such as switches and routers, both physical and virtual (hypervisor-based).

Since the number of users has grown tremendously in recent past, platforms such as social media, e-commerce, search engines, sensors etc. generate huge amount of data which is coined as Big Data. The differences with traditional data environments are: First, the way data is collected, aggregated and analyzed. Second, the infrastructure used to store and process Big Data. Third, the technologies applied to analyze large data sets. Big Data can acquire insights for better decision making in critical development areas such as health care, energy, economic productivity, natural disaster prediction etc. Since this data is highly sensitive to business, it is important to consider the security aspects while designing the architecture.

To analyse Big Data sets quickly and efficiently Google's MapReduce Paradigm [6] has been well established. Hadoop [33] is an open source implementation of the MapReduce Paradigm. Hadoop provides a distributed storage system called Hadoop Distributed File System (HDFS) and an analysis system called MapReduce. In general, Hadoop works by dividing the data processing into mainly three phases: map, shuffle/merge and reduce phase. Input is taken from HDFS and divided into many splits. For every split one MapTask runs and produces a list of <key, value> pairs. Map Output Files (MOF) are written to local storage. Shuffle phase deals with the transfer/sorting of MOFs to the nodes where reducer is scheduled to run. Before handing over the output to ReduceTask, the MapReduce framework sorts and groups the key-value pairs by key. Finally, each ReduceTask processes the segment and the final output is written to HDFS.

The rest of this paper is organized as follows. Section 3.2 provides the motivation to use SDN with Hadoop. In Sect. 3.3, several SDN applications are briefly described. After that, several security analysis techniques are mentioned in Sect. 3.4 and the results of our security analysis are shown in Sect. 3.5. Suggestions for future works are highlighted in Sect. 3.6 and an overall conclusion is drawn in Sect. 3.7.

3.2   Why to Use SDN for Handling Big Data?

Hadoop framework has several loopholes in terms of performance as shown by Wang et al. in [32]. First of all, there is a serialization barrier between shuffle/merge and reduce phases. ReduceTasks wait until all MapTasks are executed and the Map Output Files (MOF) are available. At every ReduceTask, when the total data size is greater than the memory threshold, smaller data sets are merged. Reduce function starts only when the merge/shuffle process is finished. Due to this serialization, there will be a delay in the execution of the reduce phase. Second, repetitive merge demands multiple disk access. The ReduceTask has to keep the data segments into local disk when the number of segments or their total size goes over a threshold. If more segments arrive in late, the ReduceTask has additional disk access which degrades the performance. Since Hadoop is developed in Java, it has no Remote Direct Memory Access (RDMA) support, which could lead to high speedup [32].

As SDN allows changing traffic patterns and provides access to bandwidth on demand. Researchers used these concepts to build robust Hadoop system such as FlowComb, Pythia and Hadoop-A. Since they allow the network controller to be programmable, this opens the door for intruders to perform attacks. The centralized controller is a single point of attack and failure. One possible attack is Denial of Service (DoS) where an attacker injects the network with enormous amount of packets that can exhaust system resources such as bandwidth, memory, computing power etc. This attack is the most threatening because it can paralyze the entire system. Another harmful attack is spoofing, attacker can gain access to the cluster and manipulate sensitive data by exploiting the forwarding plane. Attacker may even modify some of the code to redirect control of the traffic in such a way that it could exfiltrate data where attacker can sniff it. In this paper, several security issues in these Big Data applications are analyzed thoroughly.

3.3   SDN Big Data Applications

An overview of the following SDN applications can be seen in Table 3.1.

3.3.1   FlowComb

FlowComb [5] helps Hadoop to improve its job processing and eradicates clogging in the network by predicting network transfers before the scheduling starts and redirects the traffic to paths where sufficient bandwidth is available. Figure 3.1 shows the main components of FlowComb.

Table 3.1 Overview of the SDN applications for Big Data

SDN application	Controller used	Functionality
FlowComb [5]	NOX	Centralized decision engine finds alternative path
Pythia [20]	OpenDaylight	Uses a middleware to schedule flow
Hadoop-A [32]	RDMA	Schedules flow to improve performance

Fig. 3.1 Architecture of
FlowComb

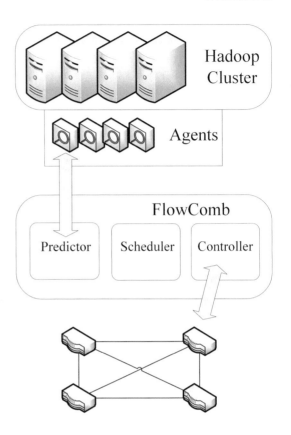

Predictor: FlowComb is equipped with software agents which are installed on each server in the Hadoop cluster. They basically scan the Hadoop logs to find out which MapTask has already finished and which transfers have been started. Then, it sends this information periodically to the Flow Scheduling module.

Scheduler: FlowComb's Scheduler detects if any of the current or pending transfer is clogging the network on their default path and schedules them to a new path.

Controller: FlowComb's Controller constructs the link to program the switches. It maintains a map of the network with all switches and the paths along with their current flow.

3.3.2 Pythia

Pythia [20] is a system that employs real-time communication intent prediction for Hadoop and uses this predictive knowledge to optimize the data center network at runtime, aiming at accelerating Hadoop's MapReduce.

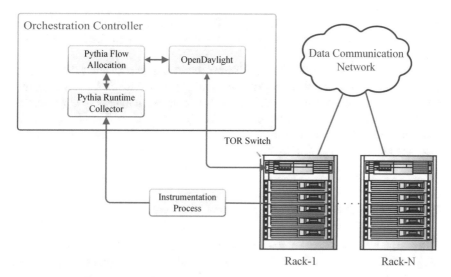

Fig. 3.2 Architecture of Pythia

Figure 3.2 shows the architecture of Pythia. It has two components:

(a) Hadoop instrumentation middleware which predicts future shuffle transfers at the level of mapper/reducer server pair during MapReduce runtime and
(b) an orchestration entity that ingests—on a per job basis—future shuffle communication intent events and optimizes the network during runtime, aiming at reducing total job completion time. The Hadoop cluster is deployed on a set of server racks (Rack-1, Rack-n in Fig. 3.2). Intra-rack data communication takes place via ToR (Top of Rack) switches that all in-rack servers connect to and inter-rack communication is provided by data communication network. It leverages the programmability offered by SDN using OpenFlow protocol to achieve a fine-grained, timely and efficient allocation of network resources to shuffle transfers.

Initially, the instrumentation process is started at every server hosting TaskTracker. The instrumentation middleware constantly monitors its local TaskTracker for job progress activity and provides the mapping of Mapper and Reducer identification from Hadoop namespace to the network location. Hadoop delays scheduling of Reducers until few Mappers have completed mapping. After the intermediate output is generated, the instrumentation process receives notification and decodes the file containing intermediate output and calculates the size of <*key, value*> pair that corresponds to each Reducer. Then it sends this predicted shuffle size in a message together with ID of respective MapTask to the Pythia server entity. A key-value pair is a set of two linked data items: a key is a unique identifier representing an item of data, and value is the content.

Pythia network scheduling module is implemented within OpenDaylight (ODL) controller [16]. ODL obtains information about physical network topology, current link-network utilization and the application communication intention. Upon receiving this information, it computes an optimized allocation of flows to network paths, such that shuffle transfers are reduced. It then maps this logical flow to the physical topology along with forwarding rules on the switches in the network.

3.3.3 Hadoop-A

There exist two approaches named Hadoop-A. Wang et al. showed an approach in [32], which is based on the RDMA [24] protocol and the second approach from Narayan et al. in [19] uses a Floodlight controller. For our security analysis, we are taking the first architecture into consideration as it eliminates one of the loopholes in Hadoop which was discussed in Sect. 3.2.

Hadoop-A achieves its speedup by utilizing RDMA-capable interconnects over Quad Data Rate (QDR) infiniBand and alternates data merge algorithms while retaining the existing Hadoop user interface. QDR infiniband is a computer-networking communication standard used in high performance computing.

It consists of two plugin components, MOFSupplier and NetMerger which are configurable and are depicted in Fig. 3.3. Both are multi-threaded C++ implementations which provide RDMA-capable interconnects and enable improved data merge algorithm. The choice of C++ instead of Java is to provide flexibility of enabling RDMA connection mechanism which Java does not support and to avoid overhead of Java Virtual Machine (JVM). MOFSupplier is equipped with a RDMA Server which takes care of the fetch requests coming from ReduceTasks. It also has a Data Engine which does the indexing and deals with the data files for all MOFs that are generated by local MapTasks.

Hadoop-A provides event channels between MOFSupplier/NetMerger and Hadoop in order to synchronize Java components. They also help in coordinating activities and monitor progress of both components. Run-time progress reports and execution statistics are stored in the Hadoop log files which help in monitoring and troubleshooting execution of Hadoop jobs.

The Hadoop-A shuffling protocol comprises of RDMA Server in MOFSupplier and RDMA Client in NetMerger. Connections are established between RDMA Server and Client using InfiniBand Reliable Connected services. Once connection is established, data is transferred through pre-registered memory buffers. The Data Engine in MOFSupplier module always prefetches data segments by retrieving data directly

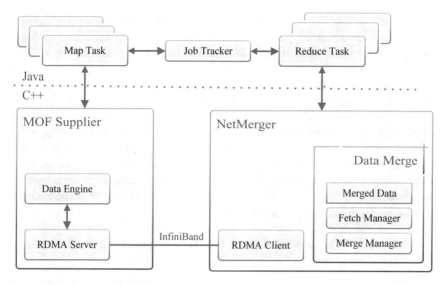

Fig. 3.3 Architecture of Hadoop-A

from the disk when they are not yet available in memory. Such data transfer is realized by a direct request and reply protocol. RDMA Client sends a request with information of designated memory buffer, and then RDMA Server finds the data and writes it to the client buffer using zero-copy RDMA write operation.

3.4   Security Analysis Methodologies

Several threat and vulnerability analysis tools and methodologies are available which can be used for security analysis such as PASTA [31], Trike [28], Attack tree [29], UMLSec [10], OCTAVE [1], Misuse Case [2], CC (Common Criteria) [18], DREAD [17], CORAS [14], and STRIDE [8].

Out of these tools and methods, we found STRIDE the most appropriate candidate for the security analysis. Some reasons against other methodologies are given in the following.

PASTA is an attack simulation methodology where user needs to know the definition and technical scope of the application and system from inside to work. The scope is limited to designers and developers. Trike is a threat modeling tool which works at the design phase as it is requirements centric and involves stakeholders. On the other hand, Attack tree, available both open source and commercial software versions, is more attacker oriented than system oriented so it does not fit when it comes to entire system analysis. UMLSec is a model based approach where each component of the system is analyzed using stereotypes which requires to know the source code. OCTAVE is a risk assessment tool which focuses on organizational risk not the technical risks. Misuse Case is a business process modeling tool based on expert guidance of various fields like architecture, design, and testing. CC is a framework for evaluating information security products which is intended for larger organizations. CC neither specifies standardization rules nor directly provides a list of product security requirements or features for specific products. DREAD is also used for risk assessment, but it was found that the ratings given to the threats are inconsistent. When it comes to risk analysis methods, CORAS is used for organizational purpose as it needs customer interaction. STRIDE deals with application level threats and categorizes them by decomposing the application into individual components, then identifies threats and suggests mitigation techniques. It can be used to perform security analysis of the application even without having any implementation of it. So it fits perfect for our analysis.

STRIDE stands for six threat categories: **S**poofing, **T**ampering, **R**epudiation, **I**nformation Disclosure, **D**enial of Service and **E**levation of Privilege. This is a threat modeling technique proposed by Microsoft.

Spoofing is impersonating an user. Proper authentication should be the security goal.

Tampering means modification of data without proper authorization. Integrity is the related security property.

Repudiation is performing malicious action without leaving a trace of it. Non-repudiation should be the target property.

Information Disclosure is the exposure of information to unauthorized users. The corresponding security property is confidentiality.

Denial of Service (DoS) is an attack to cause unavailability of services. The related property is availability.

Elevation of Privilege is the extension of user access rights. For example, if an user has 'read' access, somehow extends this to 'read-write' permission. This corresponds to the security property authorization.

In order to analyze security threats considering the aforementioned threat categories, a visualization is used which is called Data Flow Diagram (DFD) for representing different components of the system and their interactions. Table 3.2 shows the different subjects of a DFD along with their graphical representation.

Initially, we will merge different processes that are logical in the same system into one process. Each component of a system might be prone to a subset of threats. For example, data flows and data stores are prone to Tampering, Information disclosure and DoS attacks. Processes are prone to any kind of attacks. And interactors are prone to Spoofing and Repudiation.

Table 3.2 Components of DFD

Component	Symbol	Threats
Data Flow	→	Tampering, information disclosure, DoS
Data Store		Tampering, information disclosure, DoS
Process		All S.T.R.I.D.E.
Interactor		Spoofing, Repudiation
Trust Boundary		

3.5   Security Analysis of FlowComb, Pythia and Hadoop-A

The security analysis of FlowComb, Pythia and Hadoop-A is discussed in the following sections.

3.5.1   Security Analysis of FlowComb Using STRIDE

The Data Flow Diagram (DFD) of FlowComb is shown in Fig. 3.4. We merged the three parts of FlowComb into one process. We model the Hadoop instances as interactors. The FC-Agent reads the logs from Hadoop and communicates with the Flow Predictor which was merged into the FlowComb process. The Flow Controller inside FlowComb communicates with the Switches and vice versa. As the Agents run on remote machine, we put a trust boundary between the Agent and the FlowComb process.

The NOX SDN [7] controller is an open-source platform that simplifies the creation of software for controlling and monitoring networks. It supports Admission/Access control policy, Directory integration, Network monitoring and logging. The default security features of the NEC PF5240 OpenFlow switches are filter

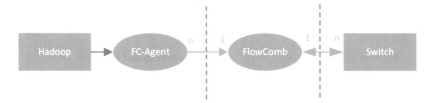

Fig. 3.4 DFD of FlowComb

Table 3.3 STRIDE Threat Matrix for FlowComb. A * indicates mitigation mechanism suggested. No threat has been mitigated natively

Type	Component	S	T	R	I	D	E
Interactor	Hadoop	*		*			
	Switch	*		*			
Data Flow	Hadoop → FC-Agent		*		*	*	
	FC-Agent → FlowComb		*		*	*	
	FlowComb ↔ Switch		*		*	*	
Process	FC-Agent	*	*	*	*	*	*
	FlowComb	*	*	*	*	*	*

(L2/IPv4/L4) and interruption of relays between ports. The security analysis of Flow-Comb using STRIDE is shown in Table 3.3.

Spoofing The FC-Agent is vulnerable to spoofing, because the Flow Predictor communicates with the FC-Agents and schedules the traffic accordingly. If Hadoop is spoofed, the adversary can access the Hadoop logs and even HDFS. As HDFS proxies are validated using IP based authentication, if somebody impersonates the user, he can manipulate valuable information or may even run malicious programs on the cluster. As FlowComb modules do not come with a prevention mechanism for such attacks, we propose the use of Kerberos authentication as in [21]. The Switches can be spoofed which can be mitigated by using a SDN Controller with Authentication, Authorization and Accounting (AAA) Services as OpenDaylight does. Similar countermeasures can be used to prevent that someone can impersonate as the FC-Controller and administrates the switches.

Tampering The centralized NOX controller is one of the main point of attack. If the controller is compromised, the attacker can disrupt the data path and the whole network gets compromised as it maintains a map of the network. The attacker may get access to a node and manipulate the data. The attacker can even get access to the Master Node, retrieve the information and location of all the clusters in the network and tamper them. This can be prevented if the access to the controller is tightly controlled. There are various Public Key Infrastructure (PKI)-based authentication protocols that can be used to mitigate this threat. A PKI is a system that can create, distribute and verify digital certificates. FlowComb uses OpenFlow protocol to communicate and the data passes through the NOX controller. OpenFlow comes with an security feature that allows the use of Transport Layer Security (TLS). TLS protects data from tampering and information disclosure by encrypting the data and uses public key cryptography, which ensures a private communication between the entities.

Repudiation A compromised process or interactor can alter the network flow and possibly deny his actions. If spoofing is taken care, then the threat of repudiation can be ignored as Agent lies inside the Hadoop cluster and we have already suggested authentication mechanisms for the Hadoop cluster. The AAA-module provides a logging mechanism. Thus, the logged actions of authenticated actors are retraceable.

Information Disclosure An attacker may gain information on the network topology and settings (FlowComb ↔ Switch), Hadoop's actions and logs (Hadoop, FC-Agent, FC-Agent ↔ Flowcomb) and all these information if FlowComb is affected by information disclosure. To protect the data in transit, we suggest the use of TLS. Hadoop, FC-Agent and FlowComb should be secured by the underlying operating system, which prohibits unauthorized data access.

Denial of Service In case the controller or data nodes get offline, the performance of the Hadoop cluster will also suffer from this. This can be prevented by separating the MapReduce process from the Authentication process. Authentication can be an independent process. DoS may affect the controller, scheduling of flows and generation of a new path but the normal Hadoop operation will carry on as it is. Another mitigation strategy suggested by Rajat et al. [11] is to limit the number of packets sent to the controller. If the controller detects that it is receiving more messages than it can handle, it could install a rule on one or more switches instructing them to send messages at a lower rate. OpenFlow is vulnerable to DoS which according to [12] can be mitigated by bounding the number of requests by using access control also called as throttling.

Elevation of Privilege A compromised FC-Agent might install malicious software on the host. This can be mitigated by executing the process with least possible rights.

Without a secure network, an attacker can easily identify the targets, gain control over the network, access and attack server, modify contents in the clusters and also modify the metadata of the name node, insert malware and monitor traffic. It is extremely important to take the security measures mentioned above to avoid such damages.

3.5.2 Security Analysis of Pythia Using STRIDE

The basic concept of Pythia is similar to the idea of FlowComb, thus the DFD seems to be very similar. As the orchestration entity and the OpenDaylight Controller are closely connected, we merged them into one process as seen in Fig. 3.5. The middleware which resides on the Hadoop machines is a process beyond our trust boundary

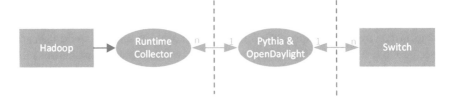

Fig. 3.5 DFD of Pythia

Table 3.4 STRIDE Threat Matrix for Pythia. A * indicates mitigation mechanism suggested and a ✓ indicates threat has been mitigated

Type	Component	S	T	R	I	D	E
Interactor	Hadoop	*		*			
	Switch	✓		✓			
Data Flow	Hadoop→ Collector		*		✓	✓	
	Collector ↔ Pythia		✓		✓	✓	
	Pythia ↔ Switch		✓		✓	✓	
Process	Runtime Collector	✓	*	✓	*	✓	✓
	Pythia	✓	*	✓	✓	✓	✓

like the FC-Agent in the previous model. The switches used are TOR switches which are OpenFlow enabled and hence use the OpenFlow protocol. Any compromise in the controller can lead to devastating consequences because the attacker is able to control the entire network. The ODL controller is one step further and established a security team to address the security issues [26]. Some of the security issues and their fixed version of ODL can be found in [3]. We have constructed a STRIDE threat matrix for Pythia as shown in Table 3.4.

Spoofing It may not be possible to spoof the OpenDaylight controller or the connected switches because the AAA service is already embedded in the controller platform [3]. AAA is implemented as a token(-claim) based authentication. User applications need this token to access controller resources, which prohibits data access by an unauthorized user. Every activity of the user is logged by an accounting feature so any malicious attempt made by the user can be easily traced. To secure the run-time Collector and the Hadoop against spoofing, we suggest a mutual authentication beween the communication parties.

Tampering Sungmin Hong et al. show how to manipulate the network topology information of ODL in [9]. They suggest using TopGuard, which is a security extension to the SDN controllers that provides automatic real time detection of Network Topology Poisoning Attacks. The data between the controller and the switches is secure as OpenFlow could be enabled with the TLS feature.

Repudiation The accounting feature of the AAA module records all access requests made by an user or an application [30]. This mitigates repudiation in ODL. In case of Hadoop, a malicious action can alter network paths. These actions will be logged as actions from the respective middleware.

Information Disclosure Vulnerabilities concerning the MD-SAL API [25] or the NETCONF protocol [4] had been closed in recent OpenDaylight releases [23]. The runtime collector has potentially sensitive data on the hadoop computations. To mitigate the risk of the runtime Collector, we advise to set the least possible permissions to the tasks and its data.

Denial of Service In ODL, the DoS threat is tackled by the application Defense4All which detects attacks against its NBI, SBI, processes, and data store as shown by Arbettu et al. in [3]. The authentication of the AAA module ensures that in case of an attack only high privileged user has access to the network resources.

Elevation of Privilege This threat could be mitigated by the AAA module.

3.5.3 Security Analysis of Hadoop-A Using STRIDE

Hadoop-A implements the RDMA protocol in order to accelerate data transfers and improves data merging. RDMA allows data transfers between the main memories of different machines without involving the processors, caches or operating systems. This leads to high-throughput and low-latency. Figure 3.6 represents the DFD of Hadoop-A. The trust boundary lies between the Hadoop cluster and RDMA clients: MOF Supplier and NetMerger. The RDMA clients lie outside of the trust boundary because they contain RDMA components which are connected to untrusted users. The results of our security analysis using STRIDE is shown in Table 3.5.

Spoofing Impersonating a legal RDMA peer is possible by spoofing a valid IP address. A network-based attacker can do this by blind attack [27] or simply establish-

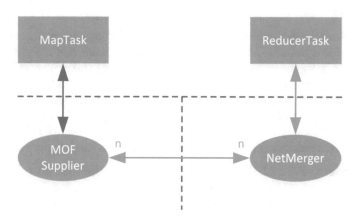

Fig. 3.6 DFD of Hadoop-A

Table 3.5 STRIDE Threat Matrix for Hadoop-A.A * indicates mitigation mechanism suggested and a ✓ indicates threat has been mitigated

Type	Component	S	T	R	I	D	E
Interactor	Map Task	*		*			
	Reducer Task	*		*			
Data Flow	Map ↔ MOF		✓		*	*	
	MOF ↔ NetMerger		*		*	*	
	NetMerger ↔ ReducerTask		✓		*	*	
Process	MOF Supplier	*	*	*	*	*	*
	NetMerger	*	*	*	*	*	*

ing an RDMA stream with the user [24]. This enables the attacker to write data into the victims memory and execute malicious code. End-to-end authentication mechanisms such as IPSec AH extension or ULP authentication prevent these attacks. In case of the interactors Map Task and Reducer Task, this threat would mean to gain write access for the corresponding areas in the main memory. This should be tackled by the operating system.

Tampering A peer can write to non-allowed memory locations using a buffer overflow. This can be used to execute malicious code. The RDMA Network Interface Card (RNIC) has to check the corresponding bounds to prevent this attack. In case of the data flow between the interactors and the RDMA clients, this should be protected by the operating system as no other process should have access to these memory regions.

Repudiation An attacker may insert malicious data into the memory of an RDMA peer. If an authentication mechanism is used, then the traffic can be logged and malicious behavior can be traced back to an RDMA client with corresponding Map/Reduce-Task.

Information Disclosure Information of the actual Hadoop computations might be accessed if this threat is not mitigated. The memory areas that are used for the DRMA service should be zeroed before advertised to mitigate the risk of information disclosure. The connection between the RDMA clients should be encrypted with IPSec ESP extension. It is possible to access buffers that were previously advertised but revoked meanwhile. The RNIC has to tackle this flaw by ensuring that only actual shared memory can be accessed. The connection between the RDMA clients and interactors resides in the same machine assumed to be sufficiently secure.

Denial of Service The RDMA protocol is susceptible to DoS attacks as a TCP SYN attack. This can stop Hadoop from working. To mitigate this threat, the allocation of resources must be done by a Privileged Resource Manager. This prevents peers from accessing more than the allocated resources.

3.6 Discussion and Future Work

Using the results of the security analysis, we are able to identify various threats and their mitigation strategies. By performing the attacks on the experimental setup, we can get concrete results and enhance the security features. The SDN applications for Big Data are still an emerging topic and require security considerations before making them commercially deployable. Although FlowComb, Pythia and Hadoop-A are similar in their application functioning but they have differences as well. Flow-Comb utilizes application domain knowledge for flow scheduling. Pythia leverages application intelligence, taking flow criticality into consideration and incorporates flow priority as a criterion in network optimization. Use of software switches in FlowComb is likely to exhibit high latency as it uses single network over subscription ratio (1:10 for 1Gbps server NICs). Whereas Hadoop-A is purely dependent on RDMA protocol over QDR infiniBand which in fact provides high-throughput and low-latency in contrast with traditional copy operation used in Hadoop. Future work will be to evaluate the performance of the suggested implementations.

3.7 Conclusion

We have analyzed FlowComb, Pythia and Hadoop-A Big Data architectures using the STRIDE method. Their commonality is that they provide performance optimizations, but security is not part of their design.

With our suggestions in Sect. 3.5, the design holes could be covered to gain security. FlowComb and Hadoop-A natively do not provide as much security services as Pythia does. Pythia uses an OpenDaylight Controller [22] which implements an Authentication, Authorization and Accounting (AAA) service, thus Pythia natively provides some security mechanisms. Since FlowComb achieves similar performance improvements as Pythia by providing less security mechanisms, we suggest to use Pythia or Hadoop-A. Pythia can speedup Hadoop by 46% while Hadoop-A doubles the data processing throughput. Implementing Pythia together with our security suggestions could be less expensive in comparison to implementing Hadoop-A with our suggestions.

References

1. Alberts CJ, Dorofee A (2002) Managing information security risks: the Octave approach. Addison-Wesley Longman Publishing Co., Inc, Boston, MA, USA
2. Alexander I (2003) Misuse cases: use cases with hostile intent. In: IEEE software20.1 (2003), pp 58-66
3. Arbettu RK (2016) Security analysis of OpenDaylight, ONOS, Rosemary and Ryu SDN controllers. In: 2016 17th International telecommunications network strategy and planning symposium (Networks). Sept. 2016, pp 37–44. https://doi.org/10.1109/NETWKS.2016.7751150

4. US-CERT/NIST. CVE-2014-5035 (2014) https://web.nvd.nist.gov/view/vuln/detail? vulnId=CVE-2014-5035 (visited on 01/04/2017)

5. Das A et al (2013) Transparent and flexible network management for BigData processing in the cloud. In: Da Silva D, Porter G (eds) 5th USENIX Workshop on Hot Topics in Cloud Computing, HotCloud'13, San Jose, CA, USA, June 25–26, 2013. USENIX Association, 2013. https://www.usenix.org/conference/hotcloud13/workshop-program/presentations/das. 56 Parvez Ahmad and Sven Jacob and Rahamatullah Khondoker

6. Dean J, Ghemawat S (2008) MapReduce: simplified data processingon large clusters. Commun. ACM 51(1):107–113. https://doi.org/10.1145/1327452.1327492

7. Gude N et al (2008) NOX: towards an operating system for networks. Comput Commun Rev 38.3:105–110. doi:https://doi.org/10.1145/1384609.1384625

8. Hernan S et al (2006) Uncover security design flaws using the STRIDE approach. In: MSDN Magazine. http://msdn.microsoft.com/en-us/magazine/cc163519.aspx

9. Hong S et al (2015) Poisoning network visibility in software-defined networks: new attacks and countermeasures. In: 22nd Annual Network and Distributed System Security Symposium, NDSS 2015, San Diego, California, USA, 8–11 Feb 2015. The Internet Society, https://www.internetsociety.org/sites/default/files/10_42.pdf

10. Jürjens J (2002) UMLsec: extending UML for secure systems development. In: International conference on the unified modeling language. Springer, New York, pp 412–425

11. Kandoi R, Antikainen M (2015) Denial-of-service attacks in Open-Flow SDN networks. In: Badonnel R et al (eds) IFIP/IEEE International symposium on integrated network management, IM 2015, Ottawa, ON, Canada, 11–15 May, 2015. IEEE, pp 1322-1326. ISBN: 978-3-901882-76-0. https://doi.org/10.1109/INM.2015.7140489

12. Klingel D et al (2014) Security analysis of software defined networking Architectures:PCE, 4D and SANE. In: Kitisin S et al (eds) Proceedings of the AINTEC 2014 on Asian internet engineering conference, Chiang Mai, Thailand, 26–28 Nov 2014. ACM, 2014, p 15. ISBN: 978- 1-4503-3251-4. https://doi.org/10.1145/2684793.2684796

13. Kreutz D, Ramos FMV, Veríssimo P (2013) Towards secure and dependable software-defined networks. In: Foster N, Sherwood R (eds) Proceedings of the SecondACM SIGCOMM workshop on hot topics in software defined networking, HotSDN 2013, The Chinese University of Hong Kong, Hong Kong, China, Friday, 16 Aug 2013. ACM, 2013, pp. 55-60. ISBN: 978-1-4503-2178-5, https://doi.org/10.1145/2491185.2491199

14. Lund MS, Solhaug B, Stølen K (2010) Model-driven risk analysis: the CORAS approach. Springer Science & Business Media

15. McKeown N et al (2008) OpenFlow: enabling innovation in campus networks. Computer Commun Rev 38(2):69–74. Security Analysis of SDN Applications for Big Data 571145/1355734.1355746. http://doi.acm.org/10.1145/1355734.1355746

16. Medved J et al (2014) OpenDaylight: towards a Model-Driven SDN Controllerarchitecture. In: Proceeding of IEEE International symposium on a Worldof Wireless, mobile and multimedia networks, WoWMoM 2014, Sydney, Australia, June 19, 2014. IEEE Computer Society, 2014, pp 1–6. ISBN: 978-1-4799-4786-DOI:10.1109/WoWMoM.2014.6918985, http://dx.doi.org/10.1109/WoWMoM.2014.6918985

17. Meier JD et al (2003) Improving web application security: threats and countermeasures. In: Microsoft Corporation 3

18. Mellado D, Fernández-Medina E, Piattini P (2007) A common criteria based security requirements engineering process for the development of secure information systems. Comput Stand Interfaces 29(2):244–253

19. Narayan D, Bailey S, Daga A (2012) Hadoop acceleration in an OpenFlow-based cluster. In: 2012 SC Companion: high performance computing, networking storage and analysis, Salt Lake City, UT, USA, 10–16 Nov 2012. IEEE Computer Society, 2012, pp 535–538. ISBN: 978-1-4673-6218-4. https://doi.org/10.1109/SC.Companion.2012.76

20. Neves MV et al (2014) Pythia: faster big data in motion through predictive software-defined network optimization at runtime. In: 2014 IEEE 28th International parallel and distributed processing symposium, Phoenix, AZ, USA, May 19–23 May 2014. IEEE Computer Society, 2014, pp 82-90. ISBN: 978-1-4799-3799-8. https://doi.org/10.1109/IPDPS.2014.20

21. O'Malley O et al (2009) Hadoop security design. In: Yahoo, Inc., Technical report
22. OpenDaylight. OpenDaylight: A Linux Foundation Collaborative Project. http://www.opendaylight.org
23. OpenDaylight. Security:Advisories. https://wiki.opendaylight.org/view/Security:Advisories#Patched_Versions_10
24. Pinkerton J, Deleganes E (2007) Direct Data Placement Protocol (DDP)/RemoteDirect Memory Access Protocol (RDMAP) Security. RFC 5042 (Proposed Standard). Internet Engineering Task Force. http://www.ietf.org/rfc/rfc5042.txt
25. OpenDaylight Project. OpenDaylight Controller: MD-SAL - OpenDaylight Project. 2017. https://wiki.opendaylight.org/view/OpenDaylight_Controller:MD-SAL. Accessed 01 April 2017
26. OpenDaylight Project. Security Advisories—OpenDaylight Project. https://wiki.opendaylight.org/view/Security_Advisories. Accessed 01 April 2017
27. Rescorla E, Korver B (2003) Guidelines for writing RFC text on security considerations. RFC 3552 (Best Current Practice). Internet Engineering Task Force, July 2003. http://www.ietf.org/rfc/rfc3552.txt. 58 Parvez Ahmad and Sven Jacob and Rahamatullah Khondoker
28. Saitta P, Larcom B, Eddington M (2005) Trike v. methodologydocument [draft]. In: http://dymaxion.org/trike/Trike_v1_Methodology_Documentdraft.pdf
29. Bruce Schneier (1999) Attack trees. Dr. Dobb's J 24(12):21–29
30. Scott-Hayward S (2005) Design and deployment of secure, robust, and resilient SDN controllers. In: Proceedings of the 2015 1st IEEE conference on network Softwarization (NetSoft), pp 1–5. https://doi.org/10.1109/NETSOFT.2015.7258233
31. Ucedavélez T, Morana MM (2016) Intro to pasta. In: RiskCentricThreat modeling: process for attack simulation and threat analysis, pp 317–342
32. Wang Y et al (2011) Hadoop acceleration through network levitated merge. In: Lathrop S, Costa J, Kramer W (eds) Conference on high performance computing networking, storage and analysis, SC 2011, Seattle, WA, USA, 12–18 Nov 2011. ACM, 2011, 57:1-57:10. ISBN: 978-1-4503-0771-0. https://doi.org/10.1145/2063384.2063461
33. White T (2015) Hadoop - The definitive guide: storage and analysis at InternetScale, 4th edn, revised & updated). O'Reilly. ISBN: 978-1-491-90163-2. http://www.oreilly.de/catalog/9781491901632/index.html

Chapter 4
Security Analysis of SDN WiFi Applications

David Artmann and Rahamatullah Khondoker

Abstract Mobile devices like smartphones, tablets and laptops demand highly-available and ubiquitous wireless networks, also named as Wireless Fidelity (WiFi) or Wireless Local Area Network (WLAN). The steadily rising amount of mobile devices implies new requirements claimed by administrators of enterprise wireless networks and owners of guest WiFi spots, such as the secure management of client authentication or the ability of load balancing. This work analyzes Odin, which solves the client association problem of wireless clients and OpenWiFi, a proto-typical approach that separates authentication, access and accounting to raise the efficiency and lower the administrative effort for guest WiFi owners. Both technologies utilize SDN to regulate their objectives. This does not only bring benefits, but also implies new security aspects. Especially because SDN in WiFi is a young sector, developers need to make sure that their software ensures a proper security level. Subsequently, both technologies are evaluated by applying the threat modeling technique STRIDE. The decision on this framework is elucidated by comparing it against other possible alternatives. Our analysis reveals that both projects do not consider security at the beginning called security by design. Fortunately, Odin and OpenWiFi can be extended by suitable countermeasures to mitigate relevant threats. These are proposed in the respective subsection of their security analysis. Conclusively, optimization suggestions pertaining to both technologies are made.

Keywords Software Defined Networking (SDN) · Security · WiFi · STRIDE
Odin · OpenWiFi · Security analysis

D. Artmann (✉)
Department of Computer Science, TU Darmstadt, Darmstadt, Germany
e-mail: artmann.david@gmail.com

R. Khondoker
Fraunhofer SIT, Darmstadt, Germany
e-mail: rahamatullah.khondoker@sit.fraunhofer.de; r.khondoker@yahoo.com

© Springer International Publishing AG 2018
R. Khondoker (ed.), *SDN and NFV Security*, Lecture Notes in Networks
and Systems 30, https://doi.org/10.1007/978-3-319-71761-6_4

4.1 Introduction

With faster connection capabilities of multi-user-multiple-input and multiple-output (MU-MIMO), stronger performing hardware and steadily upcoming amount of communicating devices, wireless network infrastructures have become increasingly complex. Especially, the growth in mobile devices and upgrades of wireless connection standards like IEEE 802.11ac imply challenging requirements, such as the proper management of the association state of clients or offering comprehensible accounting and authentication of clients, which need to be accomplished by a new approach in networking.

Software Defined Networking (SDN) offers logically centralized control capabilities, an application programming interface (API) for network administrators to dynamically initialize, control, change and manage network behavior and a flow-based paradigm that is predestined for highly scalable wireless networks [10]. In detail, SDN decouples the data plane from the control plane, as visualized by control- and infrastructure layer in Fig. 4.1. OpenFlow, as utilized communication protocol between these layers exploits the fact that most Ethernet switches share a common set of functions. OpenFlow offers an interface to program the flow table in different switches and routers to dynamically add, edit and remove entries in the flow table [15]. The application layer residing on top of the architecture, consists of software that uses SDN communication services and utilizes an interface to the control layer via the northbound API. For example, the Odin Master, which will be introduced and explained as a central instance of the Odin Framework in Sect. 4.3, is such an application. The control plane is responsible for configuring the switch and routes, while the data plane forwards packets according to the decision of the control plane. Centralized instances, hosting the control layer logic, are called SDN controllers and take care of network route computation, configuration of network devices and

Fig. 4.1 Architecture of software defined networking (based on [14])

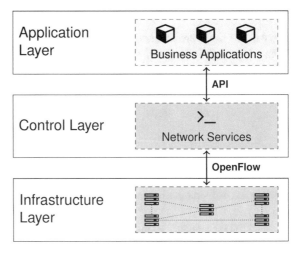

management of access control. Network elements (such as switches, routers) capable of packet switching and forwarding reside in the infrastructure layer.

Although the usage of this technology enables innovative ways in networking, it also implicates security considerations which should be taken into account. If an attacker is able to get control of a SDN controller, he or she is also able to reconfigure the whole network, respectively all components that rely on the accurate functionality of the controller instance.

This paper analyses the security capabilities of two selected technologies that utilize SDN in their architecture to determine if they implement a proper security level. For this purpose, different threat modeling frameworks are introduced in the next section and compared. This validates the most qualified approach which is subsequently used for the security analysis of Odin and OpenWiFi.

The remainder of this work is organized as follows. To justify our decision of using STRIDE, several possible security analysis frameworks are presented and compared in Sect. 4.2. Afterwards, Odins technology is introduced in Sect. 4.3.1 followed by its security analysis in Sect. 4.3.2. Similarly for the second technology, OpenWiFi, at first its architecture is introduced in Sect. 4.4.1 followed by the security analysis of its architecture in Sect. 4.4.2. Conclusively, the results of both security analyses are used while comparing Odin and OpenWiFi in Sect. 4.5. Lastly, the suggestion of possible optimization steps which exploit existing synergies of both technologies concludes this work.

4.2 Methodology

This section introduces STRIDE [12] which is used to analyze the security of Odin and OpenWiFi and gives reasons on its decision.

STRIDE is a threat modeling method developed by Microsoft and is a part of its Secure Development Lifecycle. By classifying threats in categories, STRIDE enables to identify vulnerabilities and threats in analyzed systems and their software. There are several other threat modeling techniques like OCTAVE [5], PASTA [1] or TRIKE [25]. While the first one is a complex and heavyweight solution which focuses on organizational risk but not on technical risk, the second is an attack simulation methodology where users need to be aware of the definition and technical scope of the respective application. This requires knowledge of the source code and therefore limits this approach to developers. Lastly, the third technique focuses on the design phase because it is a requirements-centric approach and involves stakeholders. We decided to use STRIDE because it is lightweight and focuses on technical risk analysis. It is used to analyze and evaluate the security of Odin and OpenWiFi. STRIDE is predestined for this task because it does not assume any implementation details of the software to be evaluated. Additionally, it does not depend on a risk model which would in turn hinge on aspects like operational condition, usage environment and customers needs. Thus, it fits for evaluating the security of software in a prototypical state like Odin and OpenWiFi. STRIDE is based on a threat modeling methodology and Data Flow Diagrams (DFD). STRIDE is an acronym for the listed threats in

Table 4.1 Threats and security properties [12, Fig. 3]

Threat	Security property
Spoofing	Authentication
Tampering	Integrity
Repudiation	Non-repudiation
Information disclosure	Confidentiality
Denial of Service (DoS)	Availability
Elevation of privilege	Authorization

Table 4.2 Symbols used in DFDs and their threats [12, Fig. 4]

Name	Symbol	Susceptible to
Data Flow	⟶	Tampering, Information disclosure, DoS
Data Store	══	Tampering, Information disclosure, DoS
Process	◯	Spoofing, Tampering, Repudiation, Information disclosure, DoS, Elevationofprivilege
Multi-process	◎	Spoofing, Tampering, Repudiation, Information disclosure, DoS, Elevation of privilege
Interactor	▭	Spoofing, Repudiation
Trust Boundary	- - - - -	

the following Table 4.1. Each of them is mapped by a security property which is necessary to be available to guard against these threats.

The basic procedure of STRIDE is to decompose a system in its components and show that each of the components is not susceptible to relevant threats. DFDs are used to visualize the interaction of components in the decomposed system. The diagram consists of standardized symbols which are shown in Table 4.2.

The one way arrow of a Data Flow represents data in motion e.g., over a network connection. A Data Store which is visualized by two parallel lines describes data at rest like files on the hard disk or databases. Processes as well as Multi-processes describe programs currently being executed. An Interactor, expressed by a rectangle, is used for endpoints in the system like web services, servers or people. Borders between untrusted and trusted elements are represented by Trust Boundaries.

Table 4.2 and the DFDs constitute a framework for investigating how the evaluated systems might fail. The following sections describe both, Odin and OpenWiFi. Afterwards, the respective security analysis is accomplished by analyzing the DFDs and the threat model.

4.3 Odin

This section introduces Odin as a technology which solves the client association problem (is explained in the following section) by using SDN and analyses its security with the already envisaged threat modeling technique STRIDE.

4.3.1 Technology

Odin [24] was introduced as a Hot Topic of SDN workshops in August 2012 on the Special Interest Group on Data Communication (SIGCOMM). Suresh et al. designed this technology to overcome specific problems of wireless networks. In detail, the IEEE 802.11 standard allows the client to decide which access points (AP) it wants to be associated with [26, P.29], hence the infrastructure is not aware of this. Furthermore, the dynamic, broadcast and time-varying nature of the wireless medium in combination with the association state machine at the AP requires to keep track of state information changes constantly. Lastly, not only associated, but also interfering IEEE 802.11 devices need to be managed. As a fundamental element and to gain simplicity for programmers, Odin entails Light Virtual Access Points (LVAP). A LVAP constitutes an abstraction layer to separate the association state from the physical AP by virtualising it. This moves the association decision on the side of the infrastructure, enables programmers to handle several clients connected to one AP as logically isolated and gives the illusion of possessing its own AP to every client by assigning an unique Basic Service Set Identification (BSSID).

Figure 4.2 visualizes the architecture of Odin and is explained as follows. To target fully centralized deployments with a SDN controller, Odin sets one Master as an

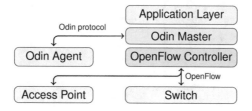

Fig. 4.2 The Master as OpenFlow application and the Odin Agents running on APs build the architecture of the Odin framework (based on [24, P.2])

OpenFlow application on top of a Floodlight SDN controller [21], communicating to Switches and APs. The Master has a global view of the network including connected clients and updates the forwarding tables on APs and Switches by using the OpenFlow protocol. Odin Agents, together with the Master implement a WiFi Split-Media Access Control (MAC) [4, P.7] which divides MAC functionalities between both parties. The Agents additionally contain the logic for LVAP handling. A Master controls Agents over a dedicated control channel via Transmission Control Protocol (TCP), which is established at boot time and allows it to add or remove LVAPs and query for statistics. Apps reside on top of the Odin Master which are allowed to inquire and examine those statistics. Each application is operated in a separate thread scheduled by the Master.

In the following, two figures and their respective explanation will contribute to a better comprehensibility of how Odin works. Given numerations in the figures will guide the reader through the process.

In Fig. 4.3, two clients are shown, Alice and Bob, connected to their exclusive LVAP. Those Light Virtual Access Points are hosted on Odin Agent 1 which runs on physical AP 1. This specific Agent as well as Agent 2, hosted on AP 2, are connected to the Odin Master via separate control channels. The use case of Figs. 4.3 and 4.4 implies a physical movement of Bobs client so that he has a stronger signal to AP 2. Going along from number 1 to 2 brings the mentioned signal strength variances on the APs. Going further along from number 2 to 3 the Odin Agent on the AP is queried for the signal strength of Bobs client by the Master. After the changed state is recognized, the Master decides to move Bobs LVAP in the custody of Agent 2.

Fig. 4.3 LVAP migration because of client movement, part I

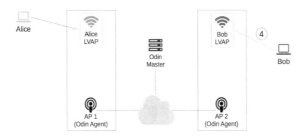

Fig. 4.4 LVAP migration because of client movement, part II

After the Master is informed as described before and is depicted in Fig. 4.3, it removes Bobs LVAP from AP 1 and adds it to AP 2, shown by number 4 of Fig. 4.4. If Bobs client has not been associated with the network before, a new LVAP on the AP which received the probe request would have been spawned, instead of moving it. Because Odin is fully transparent to the client, it does not witness this whole process.

While Sect. 4.3.1 introduced the technology of Odin the following section will document its security analysis by applying STRIDE.

4.3.2   Security Analysis

As mentioned in Sect. 4.2, a DFD is needed as for the security analysis using STRIDE. In the following, the DFD for the architecture of Odin is introduced. Afterwards, this DFD is used as a part of the threat modeling for the security analysis, targeting Odin.

Because the focus of this analysis lies on the purely Odin framework, the diagram in Fig. 4.5 entails the core of participating components: the Odin Master and an arbitrary number of Agents. Both are visualized as interactors because they are seen as end points in this model. Thus, the underlying OpenFlow controller, Access Points, and Switches as well as the Apps on top of the Master are omitted. In this figure the cardinality, as an originally foreign notation in DFDs, is introduced to point out that a Master can handle more than one Agent. Lastly, a Trust Boundary between Master and Agent reflects the inherently suspiciousness which is given by the fact that the Master has no opportunity to make sure an Agent has not been tampered with.

The adaption of STRIDE will be done from left to right with regard to Fig. 4.5, starting with the Odin Agent by stating the threats it is vulnerable to, appended with proper mitigation techniques.

Spoofing: In its present form, the implementation of the Odin Agent is not aware of a spoofed Master. This could be optimized by the use of mutual authentication, hence provide both sides with a spoofing protection like Kerberos [16].

Repudiation: In the scope of an Agent, repudiation means the inability of an Agent to proof that a specific Master has send the commands over its dedicated control channel. To avoid this lack of verifiability, both Master and Agent, should use digital signatures for signing all communicated data. As a result, an Odin Master as well

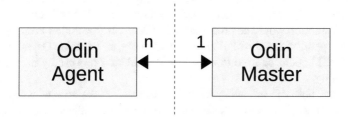

Fig. 4.5 The Data flow diagram related to Odin

as its Agents are able to validate the correctness of their counterpart. Administrators must be aware, that the introduction of digital signatures will cause a slight overhead due to the creation of signatures and their verification.

Further, the Data Flow between an Agent and its Master is analyzed.

Tampering: An attacker may manipulate the data in transit between Agent and Master. This means that the adversary is able to change a single bit in the message or add a whole payload to execute malicious code. A hindrance of this issue would be to secure the boot time established TCP/IP control channel via IPSec [13] Authentication Header on network layer or via Transport Layer Security (TLS) [7] on transport layer. Both techniques provide integrity checks, which would disclose modified packets.

Information disclosure: The data exchange between Agent and Master is not encrypted, thus an attacker is able to examine the traffic easily, because everything is sent as plaintext. An optimization regarding this lack of confidentiality is offered by IPSec. Utilizing its Encapsulating Security Payload mode, the whole traffic between Master and Agent can be encrypted. This enables to conceal the communication from any unauthorized individual and thus avoid the threat of information disclosure.

Denial of Service: This attack signifies that either a Master can not contact the Agents or vice versa. While analyzing this threat, the focus lies on the Master because it is a single point of failure and an inoperative Master would hinder the whole system to function. Whereas the unavailability of an Agent at worst causes some clients to be offline. Mitigating this attack could be done by using rate limiting or a load balancing system with high availability.

As a second interactor in the DFD of Fig. 4.5, STRIDE is also adapted on the Odin Master.

Spoofing: Although a spoofed client is prevented by the support of Wi-Fi Protected Access 2 [24, P.4], in the current implementation of Odin, a Master can not make sure that the Agent it is currently talking to is not spoofed. This could be circumvented by using a proper authentication mechanism between a Master and its Agents like the network authentication scheme RADIUS [22].

Repudiation: This threat implies that a Master can not proof, a specific Agent has communicated to it. One could claim that repudiation could be prevented by the fact, that the framework works on data link layer, i.e. with MAC addresses and thus the entity behind it can be identified. But unfortunately, MAC addresses can be spoofed. A solution would be processing each Address Resolution Protocol (ARP)-request and permitting only valid ones. Additionally, an Intrusion Detection System like Snort [23] could be used to support the validation process by monitoring for ARP-spoofing.

On a final note the following table 4.3 summarizes the antecedent analysis of the Odin framework. A X denotes that a threat could be mitigated by the suggested techniques, mentioned in the respective analysis of each threat.

Table 4.3 STRIDE threat matrix of Odin Framework

Type	Component	Threats					
		S	T	R	I	D	E
Interactors	Odin Agent	X		X			
	Odin Master	X		X			
Data flows	Agent ↔ Master		X		X	X	

A X denotes that a threat could be mitigated by the suggested techniques, mentioned in the respective analysis of each threat

4.4   OpenWiFi

After presenting the technology and the security analysis of Odin, the second part of this work introduces the architecture of OpenWiFi. Afterwards, the threat modeling technique STRIDE is applied on this prototypical technology as well.

4.4.1   Technology

Common guest WiFi systems such as CoovaChilli [6] typically are implemented as triple-A services and therefore unite access, authentication and accounting. Yap et al. introduced OpenWiFi as a prototype of separating a triple-A service into its single participating components, to reduce complexity and costs for guest WiFi owners. Additionally, the user gets relieved from the burden of remembering many different credentials for the various guest WiFi spots.

In detail, the inventors of this architecture suggest to delegate the authentication to third party service providers like Google or Facebook [27, P.4]. Those are able to handle the authentication by using techniques like OAuth2 [11] or OpenID [17]. The reason of choosing such well known authentication providers lies in their user amount, hence the probability is high that a guest WiFi user already has an account for the specific provider. Furthermore, access is provided by one or more APs which optimally support multiple SSIDs. This allows the guest WiFi provider to present several distinct WiFi networks to the user or host a private one in parallel. Additionally, if desired, the same SSID can be provided to the client across all networks. Lastly, a separate accounting service is suggested to enable responsibility delegation and billing.

The following example architecture represents the idea of OpenWiFi and is based on a given example of the inventors [27, P.5]. As in Sect. 4.3.1, a numeration is provided within Fig. 4.6 which will support the explanation.

Before going along the numeration and illuminate the given use case, each participating component deserves an introduction: going bottom up, two wireless clients, Home and Guest Device, can be seen whereby the focus lies on the latter. Surrounded by a rectangle, the guest WiFi owners OpenFlow enabled AP, a TP-Link

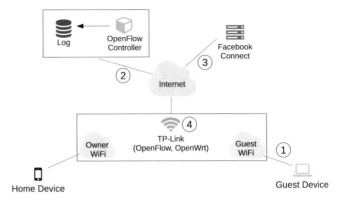

Fig. 4.6 Exemplary OpenWiFi architecture

TL-WR1043ND, hosts two SSIDs, Owner WiFi and Guest WiFi, to which the clients are connected. This AP runs on OpenWrt [18], revision backfire, which in turn runs the OpenFlow port Pantou [19] and is responsible for providing information used for access-control, redirection and accounting by the OpenFlow controller. It is enabled to offer these services, because the controller has been supplied with logic to make use of the given controls and statistics provided by the OpenFlow software switch, thus can be used as accounting service. Because Pantou seems to be a deprecated software, an exemplary technology in this case would be Open vSwitch [20]. Additionally, the OpenFlow controller logs authentication and flows to a SQLite database. Lastly, Facebook Connect is used as external authentication service provider (ASP).

After presenting the components and their roles in the OpenWiFi system, which builds on the cooperation of those, the given use case is described by going along the numeration of Fig. 4.6. Number 1 represents the initial login of a guest device to the WLAN network Guest WiFi. It is supplied with an IP address assigned via DHCP and marked as unauthenticated. While the client is labeled with this status, only its ARP, DNS and DHCP traffic is permitted [27, P.4]. Number 2, as the next step, states the redirection of a user after opening the web browser to the landing page of the accounting service. This is realized by hijacking the HTTP traffic and performing an HTTP redirect with client error code 403 [8, P.59]. In this case the user has only one choice with the mentioned authentication service (AS) of Facebook showing up on the appeared landing page. Number 3 represents clicking on the button for Facebook Connect. Subsequently, the users traffic gets forwarded to the authentication site of Facebook for entering credentials. While this happens, OpenWiFi marks the user with a new status, pending authentication. After credentials are submitted, identity is established and a valid authentication is assured, the user is asked for permission to reveal information to OpenWiFi. Confirming this dialog, the AS returns an access token to OpenWiFi and the user gets forwarded to an approval page. At this point, the user is labeled as authenticated. Lastly number 4 implies the usage of the returned access token. It enables to retrieve the users identity to associate corresponding traffic for accounting purpose.

After introducing the technology of OpenWiFi, the next Sect. 4.4.2 utilizes STRIDE to document its security analysis.

4.4.2 Security Analysis

As seen in the exemplary OpenWiFi architecture, the core components of this technology are an OpenFlow controller, a central Access Point and the external third party ASP, like Facebook Connect. Because each of the components is seen as end point in the model, these elements are visualized as interactors in Fig. 4.7. The WLAN clients as well as the network clouds are omitted since the focus lies on the core of OpenWiFi. Because each component lives on its own and does not directly belong to another, they are separated by Trust Boundaries.

With regard to Fig. 4.7, the analysis will go along the elements from left to right starting with the OpenFlow controller.

Spoofing: If an attacker is able to spoof an AP to the controller, he or she is also able to fake provided information about access control or redirection. OpenWiFi does not mitigate this threat by default. Prevention is constituted by ensuring authentication, e.g. by using certificates to sign the communicated data between Access Point and OpenFlow controller.

Repudiation: To assure non-repudiation to the controller, OpenWiFi has to implement digital signatures or the usage of timestamps to prevent deniability. If spoofing is already prevented by the usage of certificates, one can exploit this by utilizing existing certificates, assuming that they can be used for digital signatures.

Continuing with the next component, the Data Flow between OpenFlow controller and Access Point is analyzed.

Tampering: The possibility of an attacker to modify sent data between the controller and AP among others enables the attacker to tamper with accounting information and thus bypass upcoming payments. Because the OpenFlow protocol is used, which lies on top of the transport layer and uses TLS [15], the Data Flow is protected against tampering.

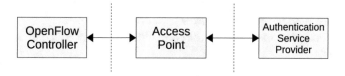

Fig. 4.7 The Data flow diagram related to OpenWiFi

Information disclosure: If TLS is used for encapsulating the traffic, any unauthorized person or program is unable to get access to the plain text behind encrypted traffic.

Denial of Service: In this context, the unavailability of a link between both parties, OpenFlow controller and Access Point, means no access control, redirection or accounting. As a consequence, the OpenWiFi architecture would be inoperative because the system can not survive without the OpenFlow controller or the Access Point. A possible mitigation, likewise described in Sect. 4.3.2, is high availability by using more than one device per side or the usage of Quality of Service by favoring the traffic between participating entities and throttling the remaining.

In the following, the Access Point as second interactor of the diagram is examined.

Spoofing: An AP is vulnerable to faked opposites in two ways. First, by a spoofed OpenFlow controller which could lead to false access controls or other unauthorized decisions. And second, in form of a spoofed authentication service, which could lead to faked identities. Both scenarios can be mitigated by the usage of mutual authentication techniques like provided by the authentication scheme Kerberos.

Repudiation: If certificates are already in use, then the administrator of OpenWiFi is able to take advantage of this by also using them for digital signatures to assure non-repudiation. If not, than additionally to certificates, the usage of timestamps would be a countermeasure for deniability.

As a fourth component, STRIDE is also adapted on the Data Flow between an Access Point and the Authentication Service Provider.

Tampering: If an adversary is able to modify the data in transit between the AP and ASP, he is also able to fake or manipulate the authentication by using another identity. This can be mitigated by the usage of HTTP over TLS (HTTPS) which uses keyed message authentication code to verify the integrity of data [7, P.13].

Information disclosure: When the connection between AP and ASP is based on HTTPS, their link is secured against exposure of any data which is transported over the channel, as long as the underlying TLS is configured with a proper cipher suite that securely encrypts the traffic.

Denial of Service: This threat needs to be split up in two parts. The first one consists of outsourcing the ASP. So in the majority of cases, the OpenWiFi architecture is unable to affect it in any way which implies its configuration and protection against DoS of any kind. Hence, OpenWiFi depends on the proper security mechanisms of the ASP and has no influence on it. It is only able to control the AP-side by offering several devices and backups for failover.

To wrap up the security analysis of OpenWiFi, we apply STRIDE on the ASPs interactor side.

Spoofing: As mentioned earlier, most probably, OpenWiFi does not influence the ASP. Thus, the latter must be protected from a spoofed AP on its side of the relation. Although the provider is not part of the native OpenWiFi system, a proper security

Table 4.4 STRIDE threat matrix of OpenWiFi

Type	Component	Threats					
		S	T	R	I	D	E
Interactors	OF controller	X		X			
	AP	X		X			
	ASP	X		X			
Data flows	OF controller ↔ AP		✓		✓	X	
	AP ↔ ASP		X		X	X	

A *X* denotes the suggestion of countermeasures and a ✓ states that OpenWiFi mitigates the threat by default. OF stands for OpenFlow

of this external part also contributes to a better one of OpenWiFi. A solution for both sides is the usage of a common certification authority which both sides trust. Hence, they can use the distributed certificates for the protection against spoofing.

Repudiation: If the AP and ASP are already using certificates, then they can rely on this to protect against deniability.

The following Table 4.4 closes up the security analysis of the prototypical approach OpenWiFi and gives an overview of the components and the respective threats they are susceptible for. A *X* denotes the suggestion of mitigation mechanisms and a ✓ states that OpenWiFi mitigates the threat by default.

4.5 Comparison and Conclusion

The vigorous surge of mobile devices, similar to mobiles predecessor, the telephone, has revolutionized communication. Their demand for a flexible and manageable network administrating approach has led to software like Odin and OpenWiFi.

Whereas the former is a framework which uses SDN to establish the central maintainability of wireless clients and their AP association, OpenWiFi is an approach to separate authentication, access and accounting to simplify the process of providing guest WiFi systems.

To facilitate a proper security analysis, we validated the most fitting technique in Sect. 4.2 and applied the elected candidate, STRIDE, to Odin and OpenWiFi.

Although both technologies have been invented for different purposes, they could be used to complement each other. To go into detail: with Odin already in use as a base of the wireless network, OpenWiFi can be built on top of it. This is possible because both technologies rely on an OpenFlow controller which could be used as synergy, assuming compatible hardware is used. As a result, it would be even more efficient and comfortable for the guest WiFi owner to offer different SSIDs or manage load balancing. Keeping this idea in mind, the administrator could also benefit of the merged security features which have to be realized when building up such a system.

Because of their prototypical state both technologies do not essentially involve security in their design yet. But as presented in Sects. 4.3.2 and 4.4.2, they fortunately entail opportunities to add security.

References

1. Morana MM, Ucedavelez T (2015, May) Application threat modeling
2. Bansal M, Mehlman J, Katti S, Levis P (2012) OpenRadio: a programmable wireless dataplane. HotSDN12. Helsinki, Finland
3. Baskett P, Shang Y, Zeng W, Guttersohn B (2013, March) SDNAN: Software-defined networking in ad hoc networks of smartphones. 10th consumer communications and networking conference (CCNC). Las Vegas, Nevada
4. Calhoun PR, Montemurro MP, Stanley D (2009, March) RFC 5416: control and provisioning of wireless access points (CAPWAP) protocol binding for IEEE 802.11. Network Working Group
5. Caralli RA, Stevens JF, Young LR, Wilson WR (2007, May) Introducing OCTAVE allegro: improving the information security risk assessment process. CarnegieMellon University
6. CoovaChilli Project. https://coova.github.io/CoovaChilli/. Accessed: December 2016
7. Dierks T, Rescorla E (2008, August) RFC 5246: The Transport Layer Security (TLS) protocol Version 1.2. Network Working Group
8. Fielding R, Reschke J (2014, June) RFC 7231: Hypertext Transfer Protocol (HTTP/1.1): semantics and content. Internet Engineering Task Force (IETF)
9. Gudipati A, Perry D, Li LE, Katti S (2013) SoftRAN: Software Defined Radio Access Network. HotSDN13. Hong Kong, China
10. Haleplidis E, Pentikousis K, Denazis S, Salim JH, Meyer D, Koufopavlou O (2015, January) RFC 7426: Software-Defined Networking (SDN): layers and architecture terminology. Internet Research Task Force (IRTF)
11. Hardt D (2012, October) The OAuth 2.0 authorization framework. Internet Engineering Task Force (IETF)
12. Hernan S, Lambert S, Ostwald T, Shostack A (2006) Uncover security design flaws using the STRIDE approach. Available: https://msdn.microsoft.com/magazine/msdn-magazine-issues. November 2006. Accessed: November, 11th 2016
13. Kent S, Seo K (2005, December) RFC 4301: Security Architecture for the Internet Protocol. Network Working Group
14. Kolias C, Ahlawat S, Ashton C, Cohn M, Manning S, Nathan S (2013, September) Openflow-enabled mobile and wireless networks. Open Network Foundation
15. McKeown N, Anderson T, Balakrishnan H, Parulkar G, Peterson L, Rexford J, Shenker S, Turner J (2008, March) OpenFlow: Enabling Innovation in Campus Networks. ACM SIGCOMM Comput Commun Rev 38(2):69–74. https://doi.org/10.1145/1355734.1355746
16. Neuman C, Yu T, Hartman S, Raeburn K (2008, July) RFC 4120: the Kerberos network authentication service (V5). Network Working Group
17. *OpenID Project*. https://openid.net/. Accessed: January 2017
18. OpenWrt Project. https://openwrt.org/. Accessed: December 2016
19. Pantou implementation. https://github.com/CPqD/openflow-openwrt. Accessed: December 2016
20. Pfaff B, Pettit J, Koponen T, Amidon K, Casado M, Shenker S (2009, October) Extending networking into the virtualization layer. 8th ACM Workshop on Hot Topics in Networks. New York City, NY
21. Project Floodlight. http://www.projectfloodlight.org/floodlight/. Accessed: December 2016

22. Rigney C, Willens S, Rubens AC, Simpson WA (2000, June) RFC 2865: remote authentication dial in user service (RADIUS). Network Working Group
23. Snort Project. https://snort.org/. Accessed: December 2016
24. Suresh L, Schulz-Zander J, Merz R, Feldmann A, Vazao T (2012) Towards Programmable Enterprise WLANS with Odin. Helsinki, Finland
25. Saitta P, Larcom B, Eddington M (2005, July) Trike v.1 methodology documentation. Octotrike
26. Yang L, Zerfos P, Sadot E (2005, June) RFC 4118: architecture taxonomy for control and provisioning of wireless access points (CAPWAP). The Internet Society
27. Yap KK, Yiakoumis Y, Kobayashi M, Katti S, Parulkar G, McKeown N (2011, July) Separating Authentication, Access and Accounting: A Case Study with OpenWiFi. Stanford University

Chapter 5
Security Analysis for the Middleware Assurance Substrate

Timm Lippert and Rahamatullah Khondoker

Abstract Middleware assurance substrate (MIDAS) is a state-of-the-art approach for Distributed Real-Time and Embedded (DRE) systems, which enables a Data Distribution Service (DDS) with Quality of Service (QoS) properties to provide performance guarantees in the system. MIDAS is based on the OpenFlow protocol for Software-defined Networking (SDN) by McKeown (INFOCOM Keynote Talk 17(2):30–32, 2009 [1]). This novel approach is designed for high performance and reliability of the system and has a low level and easy to use developer API to develop applications for the system. MIDAS is so far the first approach for DDS and QoS in SDN, which also uses OpenFlow. This approach can be used in security critical areas like the Internet of things (IoT) which lets multiple devices communicate with each other, like in smart homes where every electronic device (e.g. the fridge, TV) is connected with each other. This system that is responsible for a fast and secured communication needs to be reliable, trustworthy and secure. Since MIDAS is the first approach of DDS and QoS and is designed for performance, a security analysis is necessary for the architecture. With the STRIDE threat modeling approach used on MIDAS, the analysis will result in an overview of all possible threats for this approach to see its vulnerabilities and the techniques to mitigate the threats.

Keywords OpenFlow · STRIDE · MIDAS · Security Analysis

5.1 Introduction

The networks today often use components that are no longer state-of-the-art, because at the time they were deployed the components were the-state-of-the-art but over time the network increases and got extended by newer components which lead the network

T. Lippert (✉)
Department of Computer Science, TU Darmstadt, Darmstadt, Germany
e-mail: timm.lippert@gmail.com

R. Khondoker
Fraunhofer SIT, Darmstadt, Germany
e-mail: rahamatullah.khondoker@sit.fraunhofer.de ; r.khondoker@yahoo.com

© Springer International Publishing AG 2018   73
R. Khondoker (ed.), *SDN and NFV Security*, Lecture Notes in Networks
and Systems 30, https://doi.org/10.1007/978-3-319-71761-6_5

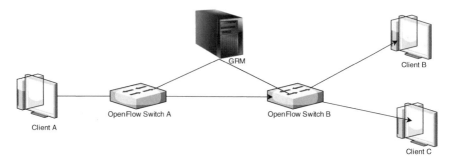

Fig. 5.1 Exemplary topology for MIDAS network

to be more and more complex. Therefore, Software-defined networking (SDN) by [1] is a novel solution by separating the network management into two different planes, the control plane and the data plane. The data plane is used for simply forwarding the data, while the control plane is responsible for defining the routes and configuration of the data plane to forward the data correctly. However, the control plane provides the applications with an abstract view on the network and enables the network control to be directly programmable.

In this case, MIDAS uses OpenFlow [2] which is the first south bound interface protocol in an SDN architecture. Through the OpenFlow protocol, MIDAS is given a remote controller to gain control over the switches in the network. In case of MIDAS, as shown in Fig. 5.1, the Global Resource Manager (GRM) which controls the OpenFlow switches allows the published data of a client to be forwarded correctly to its subscriber.

A security analysis for this approach is important because DDS is playing an important role in SDN, because of the upcoming Internet of things (IoT), which is used to connect all kind of devices with each other to exchange data [3]. For example, smart homes enable the communication of the fridge with the smart-phone to display what it contains or to allow the owner to remotely control the climate control system, or to open the garage door when the home is about to be reached. IoT is also used in health care and the automotive industry. A car for example could be connected with the phone and can show the status and location of the car. In the upcoming industrial automation called Industry 4.0 where robotic devices complete tasks with the minimum possible human interaction, in which these devices get controlled and work together over IoT. Thus IoT is expected to have a huge usage in smart home and business sector [4]. Since home and production factories are two security critical areas where security and privacy have the highest priorities and no security weaknesses are allowed (e.g. nobody would want a smart home to malfunction, or an intruder who is able to open the garage door or the main door), therefore it is necessary to have a trusted base to develop on. Hence a security analysis on MIDAS is indispensable.

The global resource manager (GRM) in MIDAS is the middleware that controls the whole traffic in the network and is responsible for the network to work as intended (e.g., who is allowed to publish data and who subscribed to which data and who can receive the data). Research for other systems that enable DDS and QoS to software-defined networks has shown that MIDAS is by now the first approach. MIDAS is also in comparison to previous approaches like real time CORBA [5] a dynamic DRE system, which allows users to simply add or subtract network components without re-configuring the system. In this paper, the architecture of the middleware assurance substrate (MIDAS) will be analyzed according to Microsoft's STRIDE threat model.

STRIDE has been chosen for the threat analysis since it focuses on application level threat modeling while other modeling frameworks like OCTAVE and Trike focus on quantifying and evaluating the identified threats [6].

An overview of the STRIDE threat modeling tool will be given in Sect. 5.2. A mandatory component in MIDAS is the OpenFlow switch, which is essential for the security analysis and the important parts for the security analysis are briefly summarized in Sect. 5.3. An example topology for MIDAS to show its functions will be described in Sect. 5.4. In Sect. 5.5, a security analysis will be conducted by using the STRIDE threat modeling on the Data Flow Diagram (DFD), which is based on the architecture of MIDAS. At last, the results from Sect. 5.5.3 will be discussed and summarized in Sect. 5.6.

5.2   STRIDE Threat Model

Microsoft's STRIDE threat modeling methodology is an effective and well used approach to find security flaws in network designs and therefore it fits perfectly on MIDAS which uses SDN [6–8]. STRIDE is an acronym and covers the following threat categories:

- **S**poofing identity:
  - Using someones identity in order to let the system think the data comes from someone else
- **T**ampering with data:
  - The data can be edited without permission and without the system noticing it
- **R**epudiation:
  - Doing something without leaving a proof that it has been done
- **I**nformation disclosure:
  - Accessing and publishing information illegitimately
- **D**enial of service:
  - Exhausting the resources of a system/service which causes the system/service to be unavailable

Table 5.1 STRIDE threat matrix shows the elements and its symbols as well as the possible threats for the elements marked with an X in the table

Threat class	Element type			
	Data flow	Process	Inter-actor	Data store
Spoofing		X	X	
Tampering	X	X		X
Repudiation		X	X	
Information disclosure	X	X		X
Denial of service	X	X		X
Elevation of privilege		X		
DFD notation	Data flow	Process	Interactor	Data store

- **E**levation of privilege:
  - Gaining more access rights than actually given by the system/administrator (e.g. an ordinary user becomes an administrator).

These threat categories are then considered for a system by transforming the system's architecture into a Data Flow Diagram (DFD). The MIDAS architecture is decomposed into its elements as shown in Table 5.1. This table also shows each element with its possible threats that help to make sure no element remains unnoticed. Once the DFD is created, every element in the system needs to be examined in terms of the suspected threats which are marked with an X in the table.

5.3   OpenFlow

Since OpenFlow switches are important components of MIDAS and have an important role in the security analysis, a quick introduction is given in this section. Figure 5.2 shows an ideal OpenFlow switch with its connected clients and the remote controller. The controller, which can be for example a server as it is in MIDAS, is connected to the switch over the OpenFlow protocol. The OpenFlow protocol encrypts the data using transport layer security protocol (TLS) to provide a secure communication between the remote controller and the OpenFlow switch [9]. This is important because the flow table that controls and forwards the traffic is managed and configured by the remote controller. This is mandatory for the security analysis in Sect. 5.5.3 because the whole system communicates over the OpenFlow switches and plays an important role for the security between the clients and the remote controller.

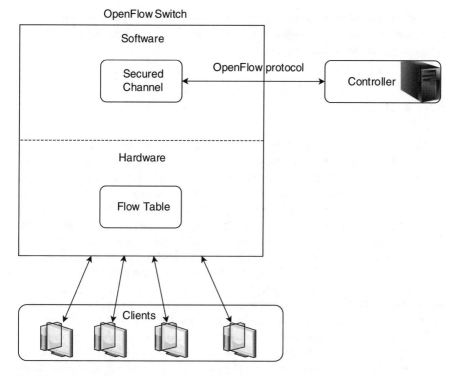

Fig. 5.2 Idealized OpenFlow switch [2]

5.4 Example Setup

The challenge described in the main security analysis in Sect. 5.5 for MIDAS is based on the exemplary topology shown in Fig. 5.1. The example topology consists of two OpenFlow switches, three clients, and a controller for the OpenFlow switches, which is the Global Resource Manager (GRM). Clients A, B and C are simple clients in the network that publish and subscribe data. For example, a publisher publishes data with the topic "IMPORTANT TOPIC", so every subscriber which is registered and has subscribed for that topic "IMPORTANT TOPIC" receives the data in that topic. The OpenFlow switches that are connected to the clients and the GRM exist to forward requests of the clients in the network. The routing is done in the GRM as soon as a switch receives a request (e.g. a publish request) from its client. The OpenFlow switch forwards the data to the GRM which then returns a valid network configuration where the data needs to be forwarded to. If the client is not allowed to publish the data, then the switch simply drops the request. The GRM stores all the information about the network such as who is allowed to publish, who did subscribe to which topic. It has an overview of the whole network including different switch specifications and determines according to the clients request, which also contains the quality of service (QoS) properties, who should receive the data and who is able to receive the data within the guaranteed time (which is submitted by the publisher) and configures the switch accordingly.

5.5 MIDAS

MIDAS [10] enables quality of service (QoS) properties for a data distribution service in distributed real-time and embedded systems [11] and allows high-performance data distribution with a publish and subscribe API for clients to publish and subscribe for data using software-defined networking. These interfaces allow the client to exchange data over the network by setting a topic to specific data or subscribe to a specific topic. A publisher also has to set the QoS properties for the publish request. Then the data, either a subscribe or publish request, is sent over the connected switch. The OpenFlow switch, which is controlled by the GRM, sends the topic, client, and QoS properties to the GRM. The GRM then checks if the request is a subscribe request and subscribes the client to the given topic or if it is a publish request the GRM checks if the client is allowed to publish data (if not the request gets dropped) and calculates all clients that subscribed to it and matches the timing guarantees which had been set in the QoS properties and configures the underlying switch properly. All clients that matched the QoS properties and subscribed to that topic then receive the data. The QoS parameters are reduced to three parameters which are the maximum latency between end-to-end user and minimum and maximum separation between updates for the topic to ensure timing guarantees in the network. According to these parameters the GRM calculates whether a subscriber is able to receive the data in a guaranteed time.

5.5.1 Challenge

Assuming that the client A wants to publish data to its subscribers client B and client C in Fig. 5.1 within a guaranteed delivery time. Client A has no information about its subscriber and simply publishes the data with a given topic which its subscribers subscribed to.

5.5.2 Solution

MIDAS solves this problem by adding QoS properties to the published data and adding a GRM as the controller to the OpenFlow switches. The GRM has an overhead over the whole network and about all subscriber and their addresses, as well as the clients that are permitted to publish data to a certain topic. Client A, which is registered to the topic in the GRM, sends its data to the OpenFlow switch with the QoS properties to ensure the timing guarantees. The OpenFlow switch sends the relevant data to the GRM which then returns the configuration for the OpenFlow switch to correctly route the data to the subscribers that match the above described QoS properties. So the OpenFlow switch simply forwards the data to the addresses chosen by the GRM.

5.5.3 Security Analysis

Figure 5.3 shows the Data Flow Diagram (DFD) of the MIDAS architecture. MIDAS is based on the OpenFlow protocol and uses OpenFlow switches with a GRM as a controller.

In the following, the elements of the DFD will be systematically analyzed and discussed for their possible threats which are listed in Table 5.1.

5.5.3.1 Data Flows

Data flow (1) needs to be protected from tampering, information disclosure and denial of service threats. An attacker with access to this data flow may be able to read and manipulate the data if it is not protected. By gaining information about the data, an attacker could gain knowledge about the data as well as the topics. Moreover, the attacker would then be able to reproduce the data to manipulate or send own fake data which would cause tampering and information disclosure. Simple countermeasures would be to encrypt the data to protect from information disclosure and integrate a message authentication code (MAC) against tampering to verify the data. One

Table 5.2 MIDAS component threat matrix. M indicates that a threat can be mitigated. T indicates that the component is susceptible to a threat

Threat class	Element type							
	Data Flow			Process		Inter-actor		Data store
	1	2, 13	3–11	1	2–5	1	2	1
Spoofing				M	M	M	M	
Tampering	M	M	M	M	M			M
Repudiation				M	M	M	M	
Information disclosure	M	M	M	M	M			M
Denial of service	M	M	M	T	M			M
Elevation of privileges				M	M			

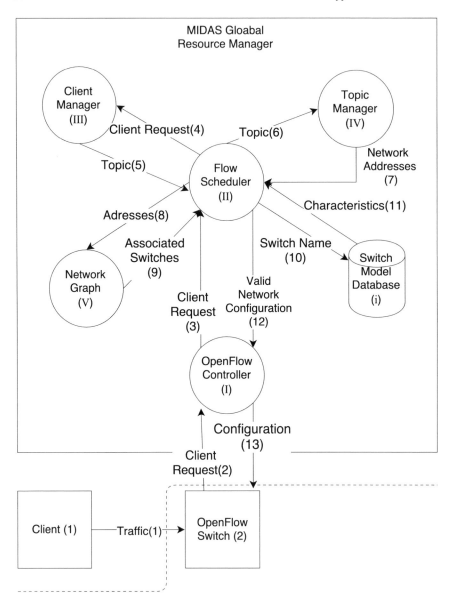

Fig. 5.3 DFD of the MIDAS approach

possible way to denial the service of data flow (1) would be to cut or congest the link that the data flow (1) uses which could be mitigated through redundant links.

Data flows (2, 13) can be considered safe against tampering and information disclosure, since the communication between the OpenFlow switch and the OpenFlow controller could be enabled using the transport layer security (TLS) as described in Sect. 5.3. Denial of service could also be mitigated by adding redundant links to data flows (2, 13) to ensure the connection, because due to structural conditions (e.g. Controller and switch can be in different buildings) the physical integrity of the links can be ensured. Additionally, since the controller which is in this case the server based GRM should be placed in a secured place (e.g. in a university, the GRM could be placed inside a server room), this physical integrity between the network components can not always be ensured because the location of the network components can be in different locations (e.g. in a campus wide network) the links established by the data flows 2 and 3 need to be redundant to mitigate the threat.

Data flow (3–12) are all located inside the GRM and have no outbound connections. Hence, they can be considered as safe. An outside attacker has no possibility to access the data for tampering or information disclosure. An outside attacker also would need physical access for a denial of service attack by cutting or intercept the links which can also be mitigated.

5.5.3.2   Processes

Processes (II–V) are all internal processes and have no connection outside the GRM and are the only processes responsible for their respective tasks and can not deny their actions nor can they come from another process. Processes (I–V) require mechanisms to protect from the DoS attack. If a single process fails, then it forces the whole system to deny its service, since all processes depend on each other. This could be mitigated by adding a process that checks all other processes and in case of a crash restarts them. Since the processes are all internal, there is no way for an attacker to gain or manipulate the information inside a process, therefore, tampering and information disclosure are unlikely.

Process (I) is the interface between the GRM and the switches. Spoofing and reputation could be mitigated by logging and digital signatures. Information disclosure and tampering can also be mitigated since the communication outside is protected by TLS. Denial of service can be mitigated by a throttling mechanism, but an attacker may be able to start a distributed denial of service attack on the GRM since an OpenFlow switch would simply forward the data to the GRM and could overload the process, which causes a real threat to the system. Throttling would in case of a DDoS attack useless, because throttling would then still cause the system to denial its service partially.

5.5.3.3 Interactors

The identity of the interactor (2) can be considered as obvious, because the switch is a main part in the system and would need to register to the GRM. Spoofing and repudiation could be mitigated since they have a static identity and their actions can be logged by the system.

The interactor (1) needs to be secured by logging its actions to prevent the repudiation threat. Spoofing could be mitigated since the interactor (1) needs to be registered to the GRM.

5.5.3.4 Data Store

Data store (i) has to be protected from tampering and information disclosure threats otherwise it could cause the flow scheduler to elect and validate the wrong routes for the configuration. The data is saved as an XML file with the information about the switches, which could easily be read and modified by an attacker. Data store (i) could be protected with a digital signature to allow only the flow scheduler to read the data, which would mitigate tampering and information disclosure. Denial of service can be mitigated, as long as the physical integrity of the GRM can be ensured, else it could be mitigated by throttling and slow down the requests.

5.6 Conclusion

The security analysis of MIDAS shows that there are several vulnerabilities in the system which can be mitigated as discussed in Sect. 5.5.3. The summary of the analysis shown in Table 5.2 shows that almost all threats can be easily mitigated with simple countermeasures. The only real threat to the system would be the distributed denial of service (DDoS) attack in which an attacker would need to compromise enough devices to manage to overload the GRM. If the GRM would be unreachable, then the whole system would be out of service. But for such an attack, an attacker would need, according to the capacity of the controller, enough compromised devices to be able to overload the GRM to be successful. A DDoS attack using QoS properties in MIDAS may not be possible since it only has properties to set a guaranteed time for delivery and no priority properties which would make the system vulnerable.

References

1. McKeown N (2009) Software-defined networking. INFOCOM Keynote Talk 17(2):30–32
2. McKeown N et al (2008) OpenFlow: enabling innovation in campus networks. ACM SIG-COMM Comput Commun Rev 38(2):69–74

3. Hakiri A et al (2015) Publish/subscribe-enabled software defined networking for efficient and scalable IoT communications. IEEE Commun Mag 53(9):48–54
4. Al-Fuqaha A et al (2015) Internet of things: a survey on enabling technologies, protocols, and applications. IEEE Commun Surv Tutor 17(4):2347–2376
5. Schmidt DC, Levine DL, Mungee S (1998) The design ofthe TAO real-time object request broker. Comput Commun 21(4):294–324
6. Markus T et al (2014) Security analysis of security applications for softwaredefined networks. In: Proceedings of the AINTEC 2014 on Asian internet engineering conference. ACM, p 23
7. Casteele SV (2004) Threat modeling for web application using STRIDE model. In: I. Royal Holloway, London
8. Shostack A (2008) Experiences threat modeling at Microsoft. In: ModelingSecurity workshop. Department of Computing, Lancaster University, UK
9. Author OpenFlow (2011) OpenFlow switch specification. Version 1:1–42
10. King AL, Chen S, Lee I (2014) The middleware assurance substrate: enabling strong real-time guarantees in open systems with openflow. In: IEEE 17th international symposium on object/component/service-oriented real-time distributed computing, 2014. IEEE, pp 133–140
11. Schmidt DC (2002) Middleware for real-time and embedded systems. Commun ACM 45(6):43–48

Chapter 6
Security Analysis of FloodLight, ZeroSDN, Beacon and POX SDN Controllers

Qamar Ilyas and Rahamatullah Khondoker

Abstract Software-defined network (SDN) is an emerging approach to replace lega-cy network's (coupled software and hardware) control and management by decou-pling the control plane (software) from the data plane (hardware). SDN provides flex-ibility to the developers by making the central control plane directly programmable. Some new challenges, such as single point of failure, might be encountered due to the central control plane. SDN focused on flexibility where as the security of the network was primarily not considered. Decoupling of control plane (software) from data plane (hardware) is a great step for innovation and research. Centralized con-trol plane may cause the single point of failure and compromising the controller means the whole network is compromised. Many organizations and data centers are moving towards SDN. Now, security is their primary concern. Security issues of the four controllers including FloodLight, ZeroSDN, Beacon and POX are analyzed with STRIDE threat modeling technique. We found that SE-FloodLight is the most secure controller because it is the most resilient controller as compared to the other controllers.

Keywords Software Defined Networking (SDN) · South Bound API (SBI) North Bound API (NBI) · STRIDE · Publish/Subscribe (pub/sub)

6.1 Introduction

SDN is a promising and emerging architectural paradigm for making the network programmable and virtualizable by decoupling control plane (software) from the data plane (hardware) [40]. In SDN, switches work as forwarding entities while the forwarding decisions are taken by the controller. These forwarding decisions are then moved to switches for the execution. It gives the programmer a control over the

Q. Ilyas
Department of Computer Science, TU Darmstadt, Darmstadt, Germany

R. Khondoker (✉)
Fraunhofer SIT, Darmstadt, Germany
e-mail: rahamatullah.khondoker@sit.fraunhofer.de; r.khondoker@yahoo.com

© Springer International Publishing AG 2018
R. Khondoker (ed.), *SDN and NFV Security*, Lecture Notes in Networks and Systems 30, https://doi.org/10.1007/978-3-319-71761-6_6

Fig. 6.1 SDN architecture
[39]

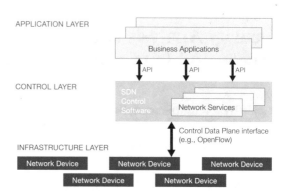

network by having the whole network image that helps in network automation and better utilization of server/network resources.

The SDN architecture can be decomposed into three layers as shown in Fig. 6.1. Data plane is also known as forwarding plane. Data plane forwards the traffic to the next hop along the path [17]. The data plane consists of network devices such as switches etc., and takes the decision of packet forwarding according to the logic of control plane. It may send packets to a particular port or push it to the controller.

The control plane may have one or more SDN controllers. OpenFlow [26] is an open source protocol that enables the interaction between the network devices, such as switches etc., with the SDN controller. The first version of the OpenFlow protocol 1.0 was released in December 2009 and then v1.1, v1.2, v1.3, v1.4 and 1.5 were released till 2014. Packet-In, Packet-Out, Modify-State, Flow-Removed etc., are the messages that are used by OpenFlow. It has become the standard for SDN that supports many protocols [43], such as TCP/IP and UDP/IP. The SDN controller sends instructions to the switches for routing, traffic engineering etc. The SDN controller acts like the brain of the OpenFlow based network that orchestrates the traffic flow and makes all the decision about packet forwarding whereas switches are only forwarding devices [21]. The controller has the topology knowledge and it is responsible to configure the data plane. This is the most important layer of the SDN architecture and this layer will be focused in this paper.

Applications that communicate with the SDN controllers via application programming interface (API) is the third layer of the SDN architecture. North Bound API (NBI) e.g., REST interfaces are used to communicate between the controller and the above layer. Applications can get the abstraction of the network state by having the image of whole network and instruct the controller for specific tasks. The communication between the controllers and applications is carried by the North-bound API (NBI). The communication between the controller and its lower layer is carried by Southbound API (SBI). The controller sends the directives (Packet-Out, Modify-State etc.) to the switches and receives requests (Packet-In etc.) from them [10]. OpenFlow is the most commonly used protocol in South Bound Interface.

The original design of the network did not take the security into the consideration because they aimed only at the service delivery [48]. Consequently, many protocols (HTTP, ARP etc.) were not resilient against the attacks (HTTP Flood [8], ARP spoofing [46] etc.) and there existed vulnerabilities that can be exploited by the attackers. Though SDN has made the network more visible (having the whole network image) and flexible (making the central control plane directly programmable) but at the same time it encounters several new security challenges (single point of failure etc.) [16]. One primary goal of SDN is to manage the data plane from a central entity that is a network controller. This feature makes SDN a client server model. By making the network as a client-server model and opening multiple interfaces to the external applications open up new challenges, such as the whole functionality can be disrupted by launching Denial of Service (DoS) and Distributed Denial of Service (DDoS) attacks [48]. The SDN controller defines the policies to prevent from such attacks.

In this paper, four controllers (FloodLight, ZeroSDN, Beacon and POX) have been analyzed on the basis of their processes and communication with the other layers. These four controllers are used in industry and research area. They have different architectures and behave differently if they are attacked. A STRIDE threat matrix is created for each of the controller to analyze the security of the controllers considering the six threat categories.

The rest of the paper is organized as follows. Section 6.2 briefly describes the STRIDE and some other threat modeling techniques. Section 6.3 sheds the light on four selected controllers. Section 6.4 covers the security analysis of each of the controllers using the STRIDE threat matrix. Section 6.5 summarizes and concludes the paper.

6.2   STRIDE

STRIDE is a threat modeling technique that was developed by Microsoft. It enables us to find the security flaws of a system. STRIDE stands for **S**poofing, **T**ampering, **R**epudiation, **I**nformation Disclosure, **D**enial of Services (DoS) and **E**levation of Privileges.

Spoofing: Impersonating as some genuine user. It can be handled by proper authentication [31].
Tampering: Modification of the data without having rights. Data integrity [7] is the related security property.
Repudiation: Doing something without leaving a proof. Non-repudiation [27] is the related security property.
Information Disclosure: Disclosure of information to an unauthorized user. Confidentiality [7] is the related property.

Denial of Services: Stopping the service availability to the genuine user. Availability [7] is the related property.

Elevation of privileges: Having more rights than the originally given. It can be handled by authorization [3].

Data Flow Diagram (DFD) is used for the graphical representation of any system. First, a system is decomposed into various components, then threats are analyzed based on the internal structure and interaction with other components. There are four standard set of symbols for DFD: data flows, data stores, processes and interactors. Modeling the DFD is the most critical point in STRIDE evaluation as all the further steps are dependent on it. If there is a problem or mistake in DFD, then all further evaluation will be incorrect. Once the components and their interaction and communication model are extracted, then the STRIDE threat categories model is applied on it.

Multiple techniques can be used for modeling the threats. It includes P.A.S.T.A (Process for Attack Simulation and Threat Analysis) [45], which is introduced by Macro Morana. Mostly, PASTA is used in application development methodologies. To be able to use PASTA, users should have a good grip on internal technical details of the system and the application. Another threat modeling technique is Flexible Modeling Framework (FMF) [14] which is used for the network design safety. Users need to know source code of application to use this method. SecureUML [24] is based on the model derived from the source code of the application. Knowledge about the source code of application is required to use it. TRIKE [36] enables communication between security team members and other stakeholders by providing a consistent framework [36]. TRIKE [36] is used to ensure that risk system implies to each asset is acceptable for all the stakeholders. User should have knowledge about the requirements and all stakeholders should be involved to use TRIKE. On the other hand DREAD [23] is an assessment tool for analyzing the risks where the subjective and inconsistent ratings could lead to inapplicable results.

Since the motivation of this paper is to analyze the components of SDN controllers and interaction between them while ignoring the internal implementation details, therefore STRIDE is the best choice for this purpose (Table 6.1).

Table 6.1 Components and symbols detail in DFD

Components	Symbols
Data flow	Arrrow
Process	Circle
Interactor	Rectangle
Trust boundary	Dotted line

6.3   SDN Controllers

A brief summary of several SDN controllers is given in the following. These controllers are FloodLight, ZeroSDN, Beacon, and POX.

6.3.1   FloodLight

FloodLight [13] is a Java-based open source controller and it is licensed by Apache. It was released in Dec, 2014 and it supports OpenFlow 1.0 and OpenFlow 1.3, along with experimental support for OpenFlow 1.1, 1.2, and 1.4 [12]. It supports both OpenFlow and non-OpenFlow protocols (Extensible Messaging and Presence Protocol (XMPP) and Border Gateway Protocol (BGP) etc.) with a broad range of virtual and physical OpenFlow switches. FloodLight supports the modular programming environment that gives the developer flexibility to add new modules on top of the existing modules.

Floodlight is considered for the analysis because many organizations, such as Canonical, SRI International, Caltech/Cern, Radware, Firemon and 6Wind, use it. The OpenDayLight (ODL) controller which is widely used is based on this controller. Therefore, any security vulnerabilities of this controller could be inherited in the ODL controller. It offers developer the ability to easily develop applications because it supports modular programming. REST APIs are included to simplify application interfaces that make it suitable to use for future enhancements.

6.3.2   ZeroSDN

SDN treats switches as "dumb" forwarding entities that is controlled by a central control logic [10]. ZeroSDN [10] brings back control onto the switch as well as benefits from the central control logic. ZeroSDN architecture splits control logic into lightweight control modules, called controllets, based on a micro-kernel approach [10]. Each controllet is implemented in a separate process and communication with other controllets and switches is carried out with messages [10]. Micro-kernel provides basic functions for messages including: pub/sub messages (routing and parsing) and discovery of other controllets etc. [10]. The micro-kernel is used for passing OpenFlow messages. These controllets are connected by a message bus that supports pub/sub communication.

It is based on communication middleware ZeroMQ [49]. ZeroSDN is an open source modular controller. It was released in 2015 [50] and supports OpenFlow 1.0 and 1.3. ZeroSDN supports different languages like python, Java, or C ++ that gives the flexiblity to the developers. ZeroSDN is considered due to its unique micro-kernel distributed architecture [10].

6.3.3 Beacon

Beacon [11] is a fast, cross-platform, modular, Java-based OpenFlow controller that supports both event-based and threaded operation [11]. Beacon was released in early 2010 [4] and it supports OpenFlow 1.0 [11]. Primarily it was built for linux but later other operating systems such as, Windows and MAC are also supported. It provides a framework to control network devices and the commonly needed plane functionality is provided by using Open Services Gateway initiative (OSGi) specification [28] and Equinox [11].

Beacon is considered because it was the first controller and most of the later controllers followed it. Beacon supports the runtime modularity meaning that it not only starts and stops applications while running but also it adds or removes those applications without shutting down the beacon process.

6.3.4 POX

POX is based on python and it is developed by Nicira. It was released in 2011 [32] and it supports OpenFlow 1.0 [33]. It is a successor of NOX that is a single threaded controller [2]. POX has an easier development environment because reusable sample components for topology discovery and path selection etc., are offered by POX. It has large support of existing APIs (JSON-RPC etc.). A web based GUI written in python is also provided by it which is helpful to shorten the development cycles. POX is used for SDN debugging, controller design and programming models [19]. POX does not support multi-threading, so performance is not dependent on number of switches [40]. POX is included in the study because it is widely used in research and it can detect invalid values of ARP header fields [40].

6.4 Security Analysis of SDN Controllers

As shown in Fig. 6.2, data flow between the controller and other two layers (data plane and applications) is carried out by an external channel because all of the three layers are fully decoupled. This decoupling of functionalities broadens the attacker surface. If the controller is not a monolithic controller, then there could be many security issues for data flows between the controllers like inter-controller trust and inter-domain trust when the controllers are not placed in same domains [44]. In this paper, only single controller security analysis is considered. Multiple controllers are not considered for analysis in this paper.

All three layers work independently and communication between the layers is carried out by an external medium. Some known attacks such as DoS etc. can be carried out because all layers are decoupled and the communication channels of con-

Fig. 6.2 Data Flow Diagram
(DFD) of SDN controllers

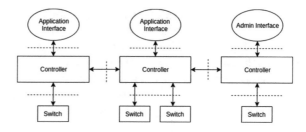

troller and other layers (Data plane and applications) may not be properly protected. Therefore, data flows will be considered to verify the confidentiality and integrity of data (information exchange). Availability of the controller will also be reasoned in the analysis.

The STRIDE matrix has been created for each controller during the analysis. In the matrix, a capital x ("X") is used when no mitigation mechanism exists for an attack. Tickmark ("✓") is used when mitigation techniques exist for an attack. In case no information is available in the research papers a minus sign ("-") is used. If a component is not affected by a threat, according to the STRIDE method, a blank space (" ") is used to denote that.

6.4.1   Floodlight

If there is no authentication service, anyone can use the network. By using password and client certificates etc., authentication is possible. Floodlight supports client certificates that is disabled by default. In NBI, Floodlight controller can be protected from spoofing attack by enabling the client certificates [22].

After receiving a request (initial connection, security assertion request, etc.) from the FloodLight controller using the REST API, a flow rule is decided in the application logic and is sent to the controller using the REST API. The communication between an application and the FloodLight controller must be protected from tampering threat. By default, HTTP is used, therefore, the communication between the controller and the application can be tampered since no integrity check mechanism is used in HTTP protocol [25]. To protect the communication from tampering in NBI, the Transport Layer Security (TLS) with HTTP (so called HTTPS) should be used by default which uses integrity check mechanisms such as SHA-256 algorithm.

The communication between the switches and FloodLight controller was not protected because FloodLight does not support TLS in SBI [37], therefore, TLS protocol should be embedded between the switches and Floodlight controller to protect the channel from tampering.

All access to REST NBI can be logged by the FloodLight controller. By default, this is disabled. To deal with repudiation in interactors (switches and applications) and processes (controller core), Floodlight can log access time, REST function and IP information. "logback.xml" file is altered to enable logging [22].

FloodLight's northbound HTTP API has no encryption and no authentication [1] that might give an intruder direct control of the FloodLight network. This makes the FloodLight controller vulnerable for Information Disclosure through the interception of the topology [41]. For NBI, disclosure of information to an unauthorized user can be protected with the TLS protocol while in SBI disclosure of information is possible because TLS is not supported in SBI.

Denial of Service (DoS) attack in NBI can be handled by introducing Rate Limiting, Event Filtering, Packet Dropping, Rule Timeout Adjustment etc. [20]. Enabling the client certificates also help to mitigate the DoS attack.

Solomon et al., [9] have conducted a distributed denial of service (DDoS) attack against Open FloodLight with user machines on the managed network, cleverly stimulating the switches to send OpenFlow "packet-in" messages to the Open FloodLight controller that consume its resources. This can be handled by restricting the PACKET OUT messages.

An already connected host may move to a new location and the controller should send the traffic to this location, this is called host tracking [51]. An intruder can impersonate as a genuine user if the HostTracker service keeps records of all connected host with MAC address. A genuine user may experience DoS or bad performance because the host migration is tracked by PACKET-IN messages and no authentication is used. This issue is addressed by FloodLight controller as MAC, IP, VLAN ID and Location is used as an index in Host Profile [15].

Authorization restricts user to use only allowed functionality. By default, Floodlight does not support authorization [22]. To deal with Elevation of privilege, Scott-Hayward introduced a system that permits the application to execute only allowed methods [38]. The newer version of the FloodLight controller called SE-FloodLight [30] has emphasized more on the security by embedding rule based authorization. This is the improvement and it has features like FortNOX [29] security that supports role based authentication using digital signatures. In case of rule conflict, the requester of high security authorization is prioritized and principle of least privilege is enforced to ensure the integrity of data. Security of Floodlight has been improved after the extended version of the FloodLight due to Authentication, Authorization and Accounting (AAA) service.

Table 6.2 has been created from the above information.

6.4.2 ZeroSDN

ZeroSDN splits the control logic into the light-weight control modules called "controllets" [10]. Latency and communication overhead can be reduced by pushing control logic to switches. Controllets are interconnected through a message bus supporting the pub/sub paradigm. Single point of failure is handled in this architec-

Table 6.2 Stride threat matrix for floodlight controller. "✓" indicates that mitigation mechanism exist for that attack, "X" indicates the possible existance of a threat but no mitigation mechanism exist, "-" indicates no information is provided for that threat, blank space indicates component can not be affected by threat

Type	Component	S	T	R	I	D	E
Data flow	North bound interface		✓		✓	✓	
Data flow	South bound interface		X		X	✓	
Process	Controller core	-	-	✓	-	✓	✓
Data store	Internal data structures		-		-	-	
Interactors	Switches and NBI applications	✓		✓			

ture [10]. Switches are decoupled from SDN controllers using middleware approach which is used successfully for communication between the services [6].

OpenFlow enabled switch is wrapped by the switch adapter (SA) to connect to message bus. Switch is connected with the SA with OpenFlow. From the switch perspective, switch adapter is its controller. SA can subscribe the OF messages (TO) that are transmitted from control applications to the switch as Packet-out etc. SA forwards only the matching messages to the switch. SA publishes (FROM) events to the bus, these are the OF messages (Packet-in etc.), that are sent from the switch to control applications [10].

There is no authentication mechanism (certificates etc.) introduced in ZeroSDN to communicate with the bus and controllets can receive all packets by applying wild card. So, spoofing is possible for any controllet in NBI.

In NBI, the channel between the controllets and bus is not secure because no encryption (SHA-256 algorithm etc.) is applied on this channel. In SBI, the channel is not secure (no encryption algorithm is applied) between the SA and bus that makes tampering and information disclosure realizable.

DoS is tackled from NBI because the controllet can only publish or subscribe the interested events and controllet has no control on the bus. DoS may not be possible from the SBI because of its architecture. In case of flooding the PACKET-IN messages, the SA disseminates such events in roundrobin [42] fashion. The bus can follow the "at most once" delivery semantic to deal with the packet flooding [10].

Repudiation might not be possible from NBI and controller core, because publish and subscription records are saved on the bus. Elevation of privilege may be possible because the user can subscribe to any messages. There is no authorization mechanism, such as Access control list (ACL), in the ZeroSDN.

The STRIDE matrix has been created in Table 6.3.

Table 6.3 Stride threat matrix for "ZeroSDN" controller. "✓" indicates that mitigation mechanism exist for that attack, "X" indicates the possible existance of a threat but no mitigation mechanism exist, "-" indicates no information is provided for that threat, blank space indicates component can not be affected by threat

Type	Component	S	T	R	I	D	E
Data flow	North bound interface		X		X	✓	
Data flow	South bound interface		X		X	✓	
Process	Controller core	-	-	✓	-	✓	X
Data store	Internal data structures		-			-	-
Interactors	Switches and NBI applications	X		✓			

6.4.3   Beacon

The Beacon OpenFlow controller [11] enables security by slicing the network control into independent virtual machines. OpenFlow applications govern the different network domains and do not interfere with other domains. In this sense, security of OpenFlow is considered as non-interference property [29].

Spoofing is very likely in Beacon in NBI. Applications are able to send any message to the controller and there is no authentication (certificates etc.) for messages between controller and applications [11].

Application Interface for Beacon is simple and impose no restrictions [11]. Applications register to receive the events. Tampering and information disclosure may be possible from the NBI because there is no encryption (SHA-256 etc.) used between applications and controller communications. In case of host tracking, tampering may not be possible because MAC, IP, VLAN ID and Location are used as an index in Host Profile [15]. The internal data structure might be vulnerable to tampering and information disclosure because once the user has access to the controller, all internal information is visible and alterable.

Confidentiality and Integrity can be achieved by using the TLS protocol in SBI that makes the channel safe against tampering and information disclosure [5]. By default, TLS is disabled in Beacon.

In NBI, conflicting flow rules from different applications can cause the DoS attack [47] while DoS attack can be launched from SBI by setting up a number of new and unknown flows [47].

There is no auditing service for applications that makes Beacon vulnerable to repudiation in NBI and processes.

Elevation of privileges may not be possible due to its slicing architecture, every application has its own domain and can not interfere with the domain of other applications. There is no access control mechanism in Beacon so elevation of privileges might be possible once the user has access to the controller.

Table 6.4 shows the STRIDE matrix for the Beacon controller.

Table 6.4 Stride threat matrix for "Beacon" controller. "✓" indicates that mitigation mechanism exist for that attack, "X" indicates the possible existance of a threat but no mitigation mechanism exist, "-" indicates no information is provided for that threat, blank space indicates component can not be affected by threat

Type	Component	S	T	R	I	D	E
Data flow	North bound interface		X		X	X	
Data flow	South bound interface		✓		✓	X	
Process	Controller core	-	-	X	-	-	X
Data store	Internal data structures		X		X	-	
Interactors	Switches and NBI applications	X		X			

6.4.4 POX

The POX [34] controller is inherited from the NOX controller. It is used to explore SDN debugging, network virtualization, controller design and programming models [19]. The STRIDE threat matrix for the POX controller is derived after integrating AuthFlow [25].

AuthFlow [25] is a security enhancement of the POX controller. The mechanism is based on RADIUS authentication server [35] and IEEE 802.1X [35]. In AuthFlow, the POX controller redirects every request from the virtual routers to the authentication server. The authenticator checks the credential against RADIUS server. If the authentication is successful, then the authenticator sends a confirmation message for the POX controller via SSL and Public Key Infrastructure (PKI). AuthFlow uses the authentication credentials for performing access control with respect to the privilege level.

Spoofing is very unlikely after embedding the authentication mechanism from AuthFlow. Interactors (switches and NBI applications) and controller core may not be spoofed because unauthenticated access is not allowed. POX can detect invalid values of ARP header fields [40].

Information disclosure is likely in POX. This is performed based on the TCP setup timing. If the second connection request is faster than the first one, then a new rule was installed on connection request. This is achieved with the POX module forwarding.l3 aggregator simple [20]. The communication between the controller and switches can be secured with TLS protocol while in NBI the communication is secured by using HTTPS that secures the channel against tampering. The internal data structure is secured from tampering and Information Disclosure by access control mechanisms.

Repudiation may be viable because POX controller does not maintain logs while switches and NBI applications communicate with the POX controller.

DoS may be achieved by sending a large number of spoofed packets to the controller that installs a new flow rule for each packet resulting the overflowing of flow table. Attack can be instantiated by using forwarding.l2 learning module [20]. This can exhaust the resources of the network controller. In NBI, authentication could

Table 6.5 Stride threat matrix for pox controller. "✓" indicates that mitigation mechanism exist for that attack, "X" indicates the possible existance of a threat but no mitigation mechanism exist, "-" indicates no information is provided for that threat, blank space indicates component can not be affected by threat

Type	Component	S	T	R	I	D	E
Data flow	North bound interface		✓		✓	✓	
Data flow	South bound interface		✓		✓	✓	
Process	Controller core	✓	✓	X	-	-	✓
Data store	Internal data structures		✓		✓	-	
Interactors	Switches and NBI applications	✓		X			

prevent from the DoS attack while in SBI, introducing the Rate Limiting and Packet Dropping etc. can mitigate these threats.

Elevation of privilege threat is tackled by AuthFlow [25]. POX processes can only access the control according to their privilege level.

Table 6.5 shows the STRIDE matrix for the POX controller.

6.5   Conclusion

Controller is the most important part of an SDN architecture, therefore, it may encounter security challenges. The design of the controller evolved with the time to deal with the flexibility, performance and security issues. Four controllers: FloodLight, ZeroSDN, Beacon and POX, are analyzed in this paper and the STRIDE threat matrix is drawn for each of these controllers. FloodLight controller does not support TLS in SBI that makes the controller vulnerable against tampering and information disclosure. The ZeroSDN architecture did not focus on security aspects in their architecture. ZeroSDN offers best performance as its maximum throughput is approximately 502k messages per second [18]. Beacon may be vulnerable to attacks such as Tampering, Information Disclosure and DoS in NBI. POX is secure as compared to the other mentioned controllers after integrating AuthFlow. SE-FloodLight is also a secure controller due to its resilience against different types of attacks. Security comes at cost of performance. If strict policies are imposed, then there will be more rejections from the controllers and performance is effected automatically. Controllers that manage the balance between performance and security is considered as the best option. SDN controllers should be updated regularly to deal with security.

References

1. Abusing Software Defined Networks (Whitepaper). https://www.blackhat.com/docs/eu-14/materials/eu-14-Pickett-Abusing-Software-Defined-Networks-wp.pdf
2. Akyildiz Ian F et al (2014) A roadmap for traffic engineering in SDN-OpenFlownetworks. Comput Netw 71:1–30
3. Authorization. http://searchsoftwarequality.techtarget.com/definition/authorization. Accessed Feb 2017
4. Beacom release. https://openflow.stanford.edu/display/Beacon/Home. Accessed Feb 2017
5. Beacon TLS. https://support.breezy.com/hc/en-us/articles/216771338-Beacon-Deployment. Accessed Feb 2017
6. Chappell D (2004) Enterprise service bus. O'Reilly Media Inc
7. CIA. http://whatis.techtarget.com/definition/Confidentialityintegrity-and-availability-CIA. Accessed Feb 2017
8. Douligeris C, Mitrokotsa A (2004) DDoS attacks and defensemechanisms: classification and state-of-the-art. Comput Netw 44(5):643–666
9. Dover JM (2013) A denial of service attack against the Open Floodlight SDNcontroller
10. Dürr F et al (2016) ZeroSDN: a message bus for flexible and light-weight network control distribution in SDN. arXiv preprint arXiv:1610.04421
11. Erickson D (2013) The beacon openflow controller. In: Proceedings of the second ACM SIG-COMM workshop on hot topics in software defined networking. ACM, pp 13–18
12. Floodlight Release and of Support. https://floodlight.atlassian.net/wiki/display/floodlightcontroller/Floodlight+v1.0. Accessed Feb 2017
13. FloodLight. http://floodlight.openflowhub.org/
14. Gilliam DP, Powell JD (2002) Integrating a flexible modeling framework(FMF) with the network security assessment instrument to reduce software security risk. In: Proceedings eleventh IEEE international workshops on enabling technologies: infrastructure for collaborative enterprises, WET ICE 2002. IEEE, pp 153–158
15. Hong S et al (2015) Poisoning network visibility in software-defined networks: new attacks and countermeasures. In: NDSS
16. Hu H et al (2014) FLOWGUARD: building robust firewalls for software defined networks. In: Proceedings of the third workshop on hot topics in software defined networking. ACM, pp 97–102
17. Kanika (2017) SDN threat analysis. http://sdntutorials.com/difference-between-control-plane-and-data-plane. Accessed Feb 2017
18. Kanika. SDN threat analysis. http://zerosdn.github.io. Accessed Feb 2017
19. Khondoker R et al (2014) Feature-based comparison and selection of Software Defined Networking (SDN) controllers. In: 2014 World Congress on Computer applications 6 security analysis of FloodLight, ZeroSDN, beacon and POX SDN Controllers 101 and Information Systems (WCCAIS). IEEE, pp 1–7
20. Klöti R, Kotronis V, Smith P (2013) Openflow: A securityanalysis. In: 2013 21st IEEE International Conference on Network Protocols (ICNP). IEEE, pp 1–6
21. Krishnan RR, Figueira N (2015) Analysis of data center SDN controller architectures: technology and business impacts. In: 2015 International Conference on Computing, Networking and Communications (ICNC). IEEE, pp 104–109
22. Laan J (2015) Securing the SDN northbound interface
23. LeBlanc D. DREADful. http://blogs.msdn.com/b/davidleblanc/archive/2007/08/13/dreadful.aspx
24. Lodderstedt T, Basin D, Doser J (2002) SecureUML: A UML based modeling language for model-driven security. In: International conference on the unified modeling language. Springer, pp 426–441
25. Mattos DMF, Duarte OCMB (2016) AuthFlow: authentication and access control mechanism for software defined networking. Ann Telecommun 71(11–12):607–615

26. McKeown N et al (2008) OpenFlow: enabling innovation in campus networks. ACM SIG-COMM Comput Commun Rev 38(2):69–74
27. Non repudiation. http://searchsecurity.techtarget.com/definition/nonrepudiation. Accessed Feb 2017
28. OSGI spec. https://www.osgi.org/developer/specifications. Accessed Feb 2017
29. Porras P et al (2012) A security enforcement kernel for OpenFlow networks. In: Proceedings of the first workshop on hot topics in software defined networks. ACM, pp 121–126
30. Porras PA et al (2015) Securing the software defined network control layer. In: NDSS
31. POX release. http://searchsecurity.techtarget.com/definition/authentication. Accessed Feb 2017
32. POX release. https://openflow.stanford.edu/display/ONL/POX+Wiki. Accessed Feb 2017
33. POX versions. https://github.com/noxrepo/pox. Accessed Feb 2017
34. Poxgit. http://github.com/noxrepo/pox/
35. Radius authentication server. http://www.elektronikkompendium.de/sites/net/1409281.htm. Accessed Feb 2017
36. Saitta P, Larcom B, Eddington M (2005) Trike v. 1 methodology document [draft]. http://dymaxion.org/trike/Trike_v1_Methodology_Documentdraft.pdf
37. Samociuk D (2015) Secure communication between Openflow switches and controllers. In: AFIN 2015, p 39
38. Scott-Hayward S, Kane C, Sezer S (2014) Operationcheckpoint: Sdn application control. In: 2014 IEEE 22nd international conference on Network Protocols (ICNP). IEEE, pp 618–623
39. sdnarchiteture. https://www.sdxcentral.com/sdn/definitions/inside-sdn-architecture. Accessed Feb 2017
40. Shalimov A et al (2013) Advanced study of SDN/OpenFlow controllers. In: Proceedings of the 9th Central & Eastern European software engineering conference in Russia. ACM, p 1
41. Shin S et al (2014) Rosemary: a robust, secure, and high-performance network operating system. In: Proceedings of the 2014 ACM SIGSAC conference on computer and communications security. ACM, pp 78–89
42. Shreedhar M, Varghese G (1996) Efficient fair queuing usingdeficit round-robin. IEEE/ACM Trans Netw 4(3):375–385
43. Song H (2013) Protocol-oblivious forwarding: unleash the power of SDN through a future-proof forwarding plane. In: Proceedings of the second ACM SIGCOMM workshop on hot topics in software defined networking. ACM, pp 127–132
44. Song H. SDN Threat analysis. https://www.ietf.org/proceedings/93/slides/slides-93-sdnrg-0.pdf
45. Ucedavélez T, Morana MM. Intro to pasta. In: Risk centric threat modeling: process for attack simulation and threat analysis, pp 317–342
46. Whalen S (2001) An introduction to arp spoofing. In: Node99 [Online Document] April
47. Yan Q, Yu FR (2015) Distributed denial of service attacks in software defined networking with cloud computing. IEEE Commun Mag 53(4):52–59
48. Zaalouk A (2014) Orchsec: an orchestrator-based architecture for enhancing network-security using network monitoring and SDN control functions. In: IEEE Network Operations and Management Symposium (NOMS). IEEE, pp 1–9
49. zeromq. http://zeromq.org. Accessed Feb 2017
50. ZeroSDN release. http://zerosdn.github.io. Accessed Feb 2017
51. Zhu Z et al (2014) Centralized flat routing. In: 2014 International conference on Computing, Management and Telecommunications (ComManTel). IEEE, pp 52–57

Chapter 7
Analysis of SDN Applications for Smart Grid Infrastructures

Marco Bräuning and Rahamatullah Khondoker

Abstract Software-defined networking (SDN) has major advantages over traditional network setups. The SDN paradigm decreases management complexity of computer networks by separating the control layer from network devices, thereby centralizing management functionalities for easier administration of the network. Due to the nature of being a massively decentralized network, it will be beneficial to integrate smart grid infrastructures with SDN. Unfortunately, critical infrastructures such as smart grids are a worthwhile target for state-sponsored cyber attacks as the past has already shown. Therefore, this work will analyze the security impact of enabling smart grids with SDN by exploring different SDN attack vectors, which will have a negative impact on smart grid security. Furthermore, a security analysis of the smart grid communication layer will be conducted, which shows Distributed Denial-of-Service (DDoS) attacks to be a major security issue for smart grid. The analysis is followed by a presentation of the SDN-enabled smart grid simulation framework *DSSnet*, which allows to assess the impact of different attack scenarios on smart grid environments, such as DDoS attacks. Although impact estimation may help to mitigate financial loss related to DDoS attacks, such attacks will still have a critical impact on smart grid operability.

7.1 Introduction

Software-defined networking (SDN) drastically reduces the management complexity of classical network topologies by separating the control layer from the data layer. This allows to abstract low-level functionalities from network devices, so that such networks can be managed in a logically centralized way and configured programmatically. Therefore, heavily decentralized networks benefit the most from SDN.

M. Bräuning (✉)
Department of Computer Science, TU Darmstadt, Germany

R. Khondoker
Fraunhofer SIT, Darmstadt, Germany

© Springer International Publishing AG 2018
R. Khondoker (ed.), *SDN and NFV Security*, Lecture Notes in Networks and Systems 30, https://doi.org/10.1007/978-3-319-71761-6_7

Examples for such decentralized networks are cellular systems or connected power networks, so-called smart grids.

Since its first official definition provided by the *Energy Independence and Security Act of 2007 (EISA-2007)*, the U.S. alone has spent over 2.5 billion USD in transitioning from traditional power networks to a smart grid [17]. By introducing a communication network layer on top of a power network layer, smart grids massively increase both reliability and efficiency as well as sustainability. Transitioning from traditional networks to a smart grid is very time consuming and expensive, since every microgrid component needs to be adapted accordingly, including electric meters at consumer households. Moreover, introducing a communication layer on top of a power network layer will also open up new attack vectors, which need to be studied and addressed before the network transition is completed.

This work analyses the applications of SDN to improve the security of smart grid environments. Therefore, past and current security risks for smart grid infrastructures will be explored and discussed. Moreover, the SDN-based smart grid simulation framework *DSSnet* [5] will be explained and its capabilities to improve the security of current smart grid infrastructures will be evaluated.

The rest of this work is structured as follows: Sect. 7.2 refreshes briefly the concepts of software-defined networking, smart grids and threat modeling with STRIDE. In Sect. 7.3, the current threat landscape and security challenges of SDN-enabled smart grids will be analyzed. In Sect. 7.4, the SDN-based simulation framework *DSSnet* will be explained. In Sect. 7.5, related and future work will be presented, followed by a conclusion of this work in Sect. 7.6.

7.2 Background

The following subsections briefly describe the concepts of smart grid infrastructures, software-defined networking and the threat modeling framework STRIDE.

7.2.1 Smart Grids

Power networks are very complex systems. In particular, they consist of power generation plants, transmission grids, distribution systems, storage units and power consumers (Fig. 7.1).

A major disadvantage of traditional power networks is the static flow of electricity, which cannot be modified according to consumer requirements. Traditional power networks are designed so that their capacity is capable of handling the maximum amount of power consumption at peak hours. During non-peak hours, the available network capacity is not used to its full extend.

Fig. 7.1 Traditional power
network architecture [8]

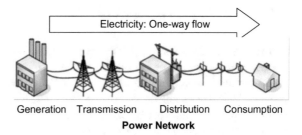

Fig. 7.2 Smart grid
architecture [8]

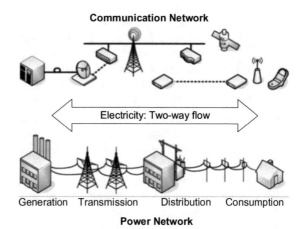

In contrast to traditional power networks, smart grids are able to control the flow of electricity, which allows to dynamically allocate power resources to consumers and therewith make a more efficient usage of the available capacity (Fig. 7.2). Basically, a smart grid is a traditional power network with an additional communication layer on-top, which enables power network components to distribute available energy as required.

Transitioning from a traditional power network to a smart grid is very cost- and time-intensive, since every network component needs to be adapted accordingly, including electrical meters at consumer households. But, once the smart grid is fully deployed, it has major benefits in contrast to traditional power networks. Dynamic allocation of electrical power will massively increase both reliability and efficiency as well as sustainability, with a major influence on current energy markets. For example, fees for power consumption could be adapted to the currently available network capacity—during peak hours, the price per kW of power will be higher than during non-peak hours.

7.2.2 Software-Defined Networking

Software-defined networking (SDN) is an approach to decrease the complexity of computer networks. After the release of Java in 1995, AT&T's *GeoPlex* [21] was the first major research project to apply the SDN paradigm to network management systems. With the establishment of the *Open Network Foundation (ONF)* in 2011 to further promote and standardize SDN methodologies, SDN has finally got more attention in research and industrial applications.

The idea of SDN is to completely separate the data layer from the control layer, which enables network administrators to manage network services through abstraction of low-level functionalities. A SDN architecture consists of three layers known as data layer, control layer and application layer (Fig. 7.3). The data layer contains all network infrastructure devices, which are controlled by the now separated control layer over a so-called control-data-plane-interface (C-DPI). The most famous CDPI is called *OpenFlow* [12] protocol, which is also standardized by the *ONF*. The control layer itself consists of programmable network services, which are managed and configured by dedicated applications running on the application layer.

The benefits of such an architecture are versatile. Decoupling the control layer from the data layer allows network administrators to dynamically adjust traffic flows according to the network clients' requirements. Since SDN networks can be configured programmatically, this dynamic adjustment can even be automatized. Moreover, SDN allows for a logically centralized management of data layer components, even if they are deployed completely decentralized across the globe. In such a scenario, SDN also enables administrators to set up and run test configurations more easily and less cost-intensively.

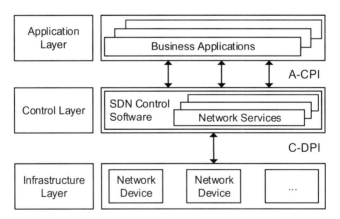

Fig. 7.3 Software-defined networking (SDN) architecture [19]

Fig. 7.4 Application of the SDN paradigm to smart grid architectures [5]

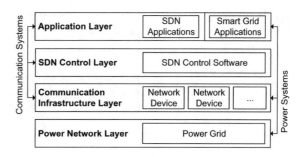

A smart grid is a use case example for SDN. Due to the decentralized nature of such power networks, SDN will decrease management complexity and therefore operational costs by a great margin. Smart grids introduced a communication layer on-top of traditional power networks, on which the SDN paradigm will be applied (Fig. 7.4).

7.2.3   STRIDE

Microsoft's *STRIDE* is a common technique for threat modeling. It can be used to assess the security of a system regarding the following threats [20]:

- **s**poofing
- **t**ampering
- **r**epudiation
- **i**nformation disclosure
- **d**enial of service
- **e**levation of privilege

Knowledge of a system's interfaces, data flows and communication protocols is required to successfully conduct a security analysis with STRIDE. Each component of the four attack surfaces data flows, processes, data storage and interactors is checked for existing mitigation mechanisms against the above listed STRIDE threat vectors. For example, if the communication protocol of a data flow does not support encryption, this component must be considered as vulnerable to information disclosure. The output of a STRIDE security analysis can be depicted as a matrix, which is also known as STRIDE threat matrix.

7.3   Security Analysis

Due to its strong impact on civil security, critical infrastructures have already been subject to cyber attacks in the past. A famous example is 2010's *STUXNET* attack,

Fig. 7.5 Attack vectors of
SDN networks [7]

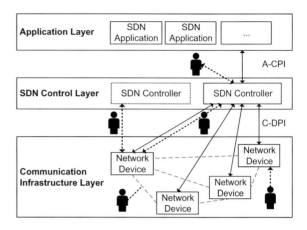

with the goal to destroy a major part of Iran's nuclear centrifuges [10]. In 2015,
Kaspersky Lab revealed similarities of the *STUXNET* code with other malware
developed from a threat actor called *Equation Group*, which can be linked to the
U.S. National Security Agency (NSA) [4]. Since such attacks on critical infrastructures are most likely politically motivated, it is also likely that they are initiated and
sponsored by a opposing government. A more recent example is the cyber attack on
the Ukrainian power grid during the Crimea crisis in December 2015. The Russian
trojan *BlackEnergy* shut down major parts of the power grid, thereby leaving 700.000
households without electricity for several hours [16]. Due to its nature, it is likely that
future power grids, such as smart grids, will continue to suffer from similar attacks.

Enabling smart grid networks with SDN has a two-fold impact on smart grid
security: On the one hand, it can be used to assess the impact of security incidents
(see Sect. 7.4) and improve resilience of the communication layer, for example by
being able to dynamically adapt network configurations in case of an attack. On the
other hand, it also introduces different attack vectors to the system [7] which are
described below and depicted in Fig. 7.5.

7.3.1 Attacks at Data Layer

New attack vectors introduced by SDN are mainly focused on the southbound API
used by the control layer to communicate with network devices on the data layer.
Since many of those protocols are very new, it is likely that they may not have been
implemented properly or hardened sufficiently against cyber attacks. A successful
compromise on this layer would allow an attacker to modify network flows, for example to bypass firewall systems or steer traffic across an attacker-controlled system to
perform a man-in-the-middle attack.

Although it should also be considered that an adversary gains unauthorized physical or virtual access to one or more of the network devices, for example to run DoS attacks against other network hosts, this is not an attack vector which results from applying the SDN paradigm.

7.3.2 Attacks at Control Layer

The SDN controller is perhaps the most obvious target of a cyber attack. A successful compromise on this layer would allow an adversary to spoof northbound and southbound API messages, thereby modifying the network flows to his advantage. Moreover, the SDN controller is the main subject of DoS attacks in an SDN environment, since its failure would have a disastrous impact on network management and network monitoring. Lastly, by setting up a rogue controller and modifying the SDN flow tables on the network devices, an attacker could initiate network flows which would not be recognized by the original controller.

7.3.3 Attacks at Application Layer

Similar to southbound protocols, these attacks leverage security vulnerabilities of the northbound API. The modus operandi of such attacks will vary depending on the API in-place (Java, Python, XML, JSON, RESTful, ...). Using STRIDE, the security of SDN-enabled smart grids can be assessed by analyzing the following data flows (see Fig. 7.4):

  i. Application Layer $\Longleftrightarrow$ SDN Control Layer
 ii. SDN Control Layer $\Longleftrightarrow$ Communication Layer
iii. Communication Layer $\Longleftrightarrow$ Power Network Layer

Data flow i and ii are directly inherited by applying the SDN paradigm to a smart grid infrastructure. Both data flows have already been analyzed for different SDN architectures and protocols in the past (see [1, 11]), with the result that DoS attacks are a major concern for SDN environments.

For the security analysis of data flow iii it will be assumed that *Open Smart Grid Protocol (OSGP)* is the protocol in use. OSGP is a communication protocol for smart meters, which was published by the *European Telecommunications Standards Institute* in 2012 [13]. With over 4 million OSGP-embedded smart meters deployed [15], it is one of the most used smart grid protocols worldwide. The result of the security analysis using STRIDE is depicted in Table 7.1.

The security analysis has shown that data flow iii has a severe impact on the security of SDN-enabled smart grids. Note that OSGP is known for its broken custom implementation of cryptography [15]. Although smart grid protocols relying on

Table 7.1 STRIDE threat matrix for data flow (iii), assuming OSGP is used. X indicates possible threat and blank indicates no applicability

Type	Component	S	T	R	I	D	E
Data Flow	Communication Layer ↔ Power Network Layer		X		X	X	

well-known cryptography standards may not suffer from tampering or information disclosure, there still would be no mitigation mechanism against DoS attacks.

7.4 DSSnet

A first approach on integrating SDN with smart grids is the SDN-based smart grid simulation framework called *Distribution System Solver Network (DSSnet)* presented by Hannon et al. in 2016. The framework integrates two core systems, the distribution power system simulator *OpenDSS* [14] and the network emulator *Mininet* [6] to offer the following features [5]:

- Studies of power flows
- Modeling of SDN-based communication networks
- Control applications for smart grid
- Emulation of virtual-time-enabled networks

The system architecture of *DSSnet* is depicted in Fig. 7.6. Besides *OpenDSS* and *Mininet*, *DSSnet* consists of three more crucial parts, in particular the network coordinator, power coordinator and a virtual time system. The two coordinators act

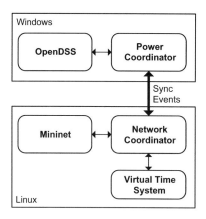

Fig. 7.6 System architecture of DSSnet [5]

as a communication middle layer between *OpenDSS* and *Mininet* for time event synchronization.

Proper synchronization was one of the key challenges for *DSSnet*. Simulators like *OpenDSS* usually use their own virtual time clock, whereas emulators like *Mininet* use the system clock as their reference point. The authors addressed this issue by developing a dedicated virtual time system for the emulator component, which is able to be synchronized with the virtual time system of the simulator component. The synchronization itself is done by freezing and unfreezing the respective system processes, which introduces further time delays known as emulation overhead. Further investigation has shown that as the number of simulated network hosts increases, the emulation overhead is also increasing (Fig. 7.7). Since smart grids are very time-critical infrastructures, this behavior is undesirable. The authors are confident to decrease the emulation overhead in a future work.

For assessing the capabilities of *DSSnet* to evaluate the impact of security incidents on smart grid infrastructures, the authors have analyzed a load shift scenario. Load shifting is the problem of properly shifting energy loads during peak hours for cost minimization. In the scenario depicted in Fig. 7.8, deploying a proper load shifting algorithm will decrease total costs from \$713.66 to \$666.01.

A key benefit of *DSSnet* is the ability to analyze real attack scenarios using the network emulation component. In the example of Fig. 7.8, a DoS attack was run against the power application server by flooding the server with TCP requests for 90 min. The simulation result shows that the impact of such a DoS attack would increase the total costs by \$22.56 to \$688.57.

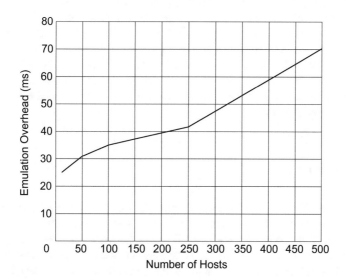

Fig. 7.7   Emulation overhead in DSSnet [5]

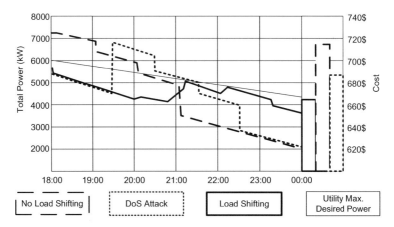

Fig. 7.8 Results of a load shifting simulation with DSSnet [5]. A simulated DoS attack starts at 19:30 and lasts for 90 min

7.5 Related and Future Work

Cahn et al. [2] did a first approach on enabling smart grids with SDN by designing and deploying a substation network architecture using a Ryu-based SDN controller. In 2014, Fujitsu Network Communications Inc. published a tech report on the challenges of SDN for the energy and utilities sector [19]. Dorsch et al. [3] also analyzed applications of the SDN paradigm to smart grids, but not from a security point-of-view. They presented a dynamic network control approach based on SDN for smart grids and evaluated their work using multiple failure scenarios, such as link congestion and corresponding recovery solutions. Dong et al. [22] discuss the opportunities that SDN brings to smart grids for improving resilience and the corresponding challenges that still remain. Kim et al. [9] investigated the state-of-the-art of SDN-enabled smart grids and compared recent studies on the service functionalities of SDN-enabled smart grids.

7.6 Conclusion

Smart grids have many benefits over traditional power networks and therefore it is only a matter of time until the first smart grids will be fully deployed. Nevertheless, adding a communication layer on-top of the power network has negative implications on security, since it opens up new attack vectors for adversaries [18]. Due to being a critical infrastructure, smart grids will continue to be a worthwhile target for cyber attacks, as it is already the case for traditional power networks today.

Applying the SDN paradigm to a smart grid will have both positive and negative impacts on its security. Due to the centralized management capabilities, monitoring and incident response are expected to be more efficient. In case of an incident, network administrators can quickly disable, enable or change affected network flows to mitigate the impact of the attack. Moreover, SDN's traffic simulation capabilities can be used to assess different attack scenarios on various network setups prior to their deployment in production.

Apart from that, SDN also introduces several attack vectors to smart grids, which need to be addressed. As this work has shown, a major issue for smart grid infrastructures are DoS attacks. Since this is also true for SDN infrastructures, SDN-enabled smart grids will continue to suffer from this threat vector. The SDN-based simulation framework *DSSnet* has shown that by being able to simulate the behavior of network traffic under different network configurations in advance, the financial impact of DoS attacks can be estimated. Although such simulations can be used to optimize network configuration for cost minimization, DoS attacks will continue to have a major financial impact. Therefore, addressing this attack vector is still an open research question.

References

1. Bayarou K, Arbettu RK, Khondoker R, Weber F (2016) Security analysis of OpenDay-Light, ONOS, RoseMary and RYU SDN controllers. In: Networks 2016, 17th international network strategy and planning symposium. Montreal, Canada. http://publica.fraunhofer.de/dokumente/N-404694.html
2. Cahn A et al (2013) Software-defined energy communication networks: from substation automation to future smart grids. In: 2013 IEEE International conference on smart grid communications (SmartGridComm), Oct 2013. pp 558–563. https://doi.org/10.1109/SmartGridComm.2013.6688017
3. Dorsch N et al (2014) Software-defined networking for smart grid communications: applications, challenges and advantages. In: 2014 IEEE international conference on smart grid communications (SmartGridComm), Nov 2014. pp 422–427. https://doi.org/10.1109/SmartGridComm.2014.7007683
4. Equation group: questions and answers (2015) Technical Report, Kaspersky Lab, Feb 2015. https://securelist.com/files/2015/02/Equation_group_questions_and_answers.pdf
5. Hannon C, Yan J, Jin D (2016) DSSnet: a smart grid modeling platform combining electrical power distribution system simulation and software defined networking emulation. In: Proceedings of the 2016 annual ACM conference on SIGSIM principles of advanced discrete simulation, SIGSIM-PADS '16. Banff, Alberta, Canada: ACM. pp 131–142. ISBN: 978-1-4503-3742-7. https://doi.org/10.1145/2901378.2901383
6. Heller B, Lanth B, McKeown N (2010) A network in a laptop: rapid prototyping for software-defined networks. In: Proceedings of the 9th ACM SIGCOMM workshop on hot topics in networks. pp 19:1–19:6
7. Hogg S (2014) SDN security attack vectors and SDN hardening (Oct 2014). http://www.networkworld.com/article/2840273/sdn/sdn-security-attack-vectors-and-sdn-hardening.html
8. Introduction to Smart Grid (2012) Technical report, Texas Tech University. http://www.ee.ucr.edu/~hamed/Smart_Grid_Topic_2_Smart_Grid.pdf
9. Kim J, Filali F, Ko Y-B (2015) Trends and potentials of the smart grid infrastructure: from ICT sub-system to SDN-enabled smart grid architecture. Appl Sci 5(4):706–727. ISSN:2076-3417. https://doi.org/10.3390/app5040706, http://www.mdpi.com/2076-3417/5/4/706

10. Langner R (2013) To kill a centrifuge—a technical analysis of what Stuxnet's creators tried to achieve. Technical report, The Langner Group, Nov 2013. http://www.langner.com/en/wp-content/uploads/2013/11/To-kill-a-centrifuge.pdf
11. Marx R, Brandt M, Khondoker R, Bayarou K (2014) Security analysis of software defined networking Protocols-OpenFlow, OF-Config and OVSDB. In: Fifth IEEE international conference on communications and electronics ICCE'14, Da Nang, Vietnam. http://publica.fraunhofer.de/dokumente/N-323861.html
12. McKeown N et al (2008) Openflow: enabling innovation in campus networks. Technical report, March 2008. http://archive.openflow.org/documents/openflow-wp-latest.pdf
13. Open Smart Grid Protocol (OSGP) (2012) Technical report, European Tele Communication Standard Institute, Analysis of SDN Applications for Smart Grid Infrastructures xvii. http://www.etsi.org/deliver/etsi_gs/OSG/001_099/001/01.01.01_60/gs_osg001v010101p.pdf.1
14. OpenDSS—EPRI Distribution System Simulator. Technical report. https://sourceforge.net/projects/electricdss/
15. OSGP: Foolish Crypto on the Smart Grid (2015) Technical report, SSL.com, May 2015. https://www.ssl.com/article/osgp-foolish-crypto-on-the-smart-grid/
16. Potential Sample of Malware from the Ukrainian Cyber Attack Uncovered (2016) Jan https://ics.sans.org/blog/2016/01/01/potential-sample-of-malware-from-the-ukrainiancyber-attack-uncovered
17. U.S. smart grid spending from 2008 to 2017, by segment (in billion U.S. dollars) (2014) Technical report, Bloomberg New Energy Finance; US Department of Energy, Aug 2014. https://www.statista.com/statistics/222082/projected-us-smart-grid-market-size-since-2009/
18. Smart-Grid Security Issues (2010) Technical report, IEEE computer and reliability societies, Jan 2010. https://www.computer.org/cms/Computer.org/ComputingNow/homepage/2012/0112/SG_SP_SmartGridSecurityIssues.pdf
19. Software-Defined Networking for the Utilities and Energy Sector (2014) Technical report, 2801 Telecom Parkway, Richardson, Texas: Fujitsu Network Communications Inc., Feb 2014. http://fujitsu.com/us/Images/SDN-for-Utilities.pdf
20. The Stride Threat Model Technical report, Microsoft Corporation. https://msdn.microsoft.com/en-us/library/ee823878(v=cs.20).aspx
21. Vanecek G (1997) GeoPlex: Universal Service Platform for IP Network-basedServices. Technical report AT and T, Oct 1997. http://www.cerias.purdue.edu/news_and_events/events/security_seminar/details/index/56218-noZCKZKV1F3c-372hq15nTbsPk31bQ8W
22. Xinshu D et al (2015) Software-defined networking for smart grid resilience: opportunities and challenges. In: Proceedings of the 1st ACM workshop on cyber-physical system security. CPSS '15. Republic of Singapore: ACM, Singapore, pp 61–68. ISBN: 978-1-4503-3448-8. https://doi.org/10.1145/2732198.2732203

Chapter 8
Security Analysis of SDN WAN Applications—B4 and IWAN

Rajat Jain and Rahamatullah Khondoker

Abstract Software defined applications for WAN (Wide Area Network) are primarily designed to manage and deploy enterprise WAN infrastructure. SDN controller feature helps an organization to automate complex WAN configuration and route data efficiently among its remote sites from a centralized point. Recently various vendors have stepped in this market and claim their product to be the solution for WAN management problems. However, automating the network through a centralized controller makes the network a handy target for attackers to exploit. Compromising the controller or its application can pose serious threat to network devices and traffic flow. This motivated us to study the vulnerabilities of two such SDN WAN applications—Google's B4 and Cisco's IWAN. For the analysis, we used the Microsoft's threat analysis method called STRIDE. In the analysis, we found out that both B4 and IWAN might suffer from security threats like Spoofing, Tampering, Information Disclosure and Denial of Service (DoS) and each vulnerability found in the application using STRIDE threat model, can be mitigated using available IT security mechanisms.

Keywords WAN · Google's B4 · IWAN · Microsoft's threat analysis · STRIDE
SDN WAN · Data Flow Diagram (DFD)

8.1 Introduction

Over the past few years, the rise in data volume of IP applications has been quite unprecedented. According to the study in [1], data usage will surpass zettabyte (i.e. 1000 exabyte) in 2018. In addition to sharp data rise, service level agreement and policies for mission critical application is making network management complex and challenging for an organization [1]. Many attempts have been made in the past to ease

R. Jain (✉)
Department of Computer Science, TU Darmstadt, Darmstadt, Germany
e-mail: rajat.jain@stud.tu-darmstadt.de

R. Khondoker
Fraunhofer SIT, Darmstadt, Germany
e-mail: rahamatullah.khondoker@sit.fraunhofer.de; r.khondoker@yahoo.com

© Springer International Publishing AG 2018
R. Khondoker (ed.), *SDN and NFV Security*, Lecture Notes in Networks
and Systems 30, https://doi.org/10.1007/978-3-319-71761-6_8

1. Interface status update.
2. Controller processing update.
3. API accessing update.
4. Flow entries updated
5. Routing Table updated

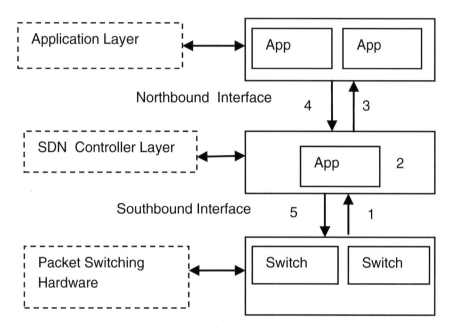

Fig. 8.1 Basic SDN Architecture and flow of data [5] (redrawn)

out this complexity but changing the underlying IP network has always been the major obstacle [2]. Whereas, SDN has proposed the idea of decoupling the control and the data plane while orchestrating the whole network through a centralized controller. With the overview of the whole topology at a single point makes the management easy and will cut down OPEX and CAPEX. Figure 8.1 shows the basic architecture of SDN. It can be viewed as three layers. For understanding the architecture in detail, the numbers are marked on the figure in increasing order to represent the flow of data, (1) The OpenFlow [3] protocol enables switches using southbound interfaces to send the network update to the controller, (2) Controller collects the updates and prepares it for the northbound applications, (3) Applications using northbound interface access the data and process it using its business logic, (4) Application forwards the instruction which updates the flow table at the controller, (5) Finally, controller using southbound interface updates the switch flow table.

Table 8.1   Various SDN solutions

SDN	Vendor	Description	Area of use	Security
IWAN [7]	Cisco	Virtual WAN, Intelligently routes based on priority, uses DMPVN cell5	WAN	Firewall (Based on security rules) and Cloud-based security applications
ION [8]	Cloud Genix	SD-WAN, virtual elements on existing device acts as flow forwarders	WAN	OpenFlow (SSL, TLS) and Cloud-based security applications
SEN [9]	Viptela	Delivers secure end-to-end virtualization for enterprise to build large scale network	WAN	Datagram TLS, IPSec
B4 [10]	Google	Connecting data centers of Google worldwide	WAN	OpenFlow (SSL, TLS) based
SDX [11]	Proto type	Software defined IXP	WAN	OpenFlow (SSL, TLS) based

The management of a WAN was always a big concern for organizations with multiple branches. Over the past years, the organizations have invested a lot in their WAN infrastructure, but still find it hard when it comes down to the management of QoS for business critical applications. Hybrid WAN allows enterprises to reduce cost by dynamically setting QoS and routing path for different classes of traffic over private and public links, but it simultaneously increases the configuration and management complexity for the network team [4].

Various SDN WAN applications have been developed by multiple vendors over the past few years. Few of those applications are listed in Table 8.1 along with their built-in security features. For our study, we decided to analyze B4 and IWAN. We have chosen B4 due to its implementation in large data center networks of Google, and IWAN because it includes features like Wide Area Application Services (WAAS), Application Visibility and Control (AVC) and zero touch deployment of Dynamic Multi-point VPN (DMPVN) make it a complete WAN management suite. The rest of the paper is structured as follows: Sect. 8.2 gives an introduction to STRIDE and DFD. Sections 8.3 and 8.4 present a summary of the B4 and IWAN applications respectively and their security analysis and Sect. 8.5 presents the conclusion.

8.2   Threat Modeling Tools

Threat modeling of applications have become the key part of product development life cycle in most organizations these days. It not only finds the potential vulnerabilities in a product but also helps the team members to understand the product in more detail [12]. Moreover, threat modeling provides a structured display of causes that could compromise the security of an application. General steps to follow for any threat modeling includes the creation of a structural overview of the application, splitting the application into its constituent components, identification of threats in each part and documentation of threats prioritizing the threats, and selection of mitigation techniques against those threats. Different schemes for identifying and classifying threats have been developed over the years. Following are some of them listed along with their alignment to the study conducted in this paper.

PASTA   Initial letters stand for Process for Attack Simulation and Threat Analysis. It is a simulation methodology suitable for designers and developers in the organization where the user needs to know the definition, technical scope and source code of application [13].

Trike   It is a threat modeling technique suitable for design phase as it is a requirement centric method and needs stakeholders involvement [14].

Attack Tree   It is available as open source as well as commercial software but it is an attacker oriented method rather than a system centric one, and therefore, may not be suitable for an entire system analysis [15].

UMLSEC   It is a model-based approach where each component of the system is analyzed with various stereotypes which requires an analyzer to know the source code of the product [16].

OCTAVE   It is a risk assessment tool for organizations where an analysis team of expertise from various departments is required for the analysis [17].

Misuse Cases   It is a business process modeling tool based on the expert guidance of various fields like architecture, design [18].

DREAD   It is also used for risk assessment, but it is more subjective in nature when one need to give ratings to the threats [19].

CORAS   It is used for the organizational threats and it needs customer interaction for the security analysis. Study in this paper does not involve any kind of customer interaction thus make it a non suitable tool for our analysis [20].

STRIDE   Comparing with above schemes STRIDE [21], a methodology from Microsoft, is perfectly suitable for analyzing our software defined wide area network applications B4 and IWAN because it does not require implementation details for categorizing and finding potential threats.

8.2.1 The STRIDE Threat Modeling Tool

STRIDE is a methodology for the identification and categorization of various threats in applications [21]. Initial letters define the security threats such as Spoofing, Tampering, Repudiation, Information disclosure, Denial of Service (DoS), and Elevation of Privileges. According to [21], individual security threat is defined as follows.

Spoofing means pretending to be an authorize user for accessing a certain service or application.

Tampering involves unsanctioned manipulation of data. Modification of data could be done during transmission or when the data is in rest.

Repudiation means denial of action after its occurrence. For example, an user denies the truth that the action was performed by him.

Information Disclosure involves the leak of information to anyone, who is not authorized to access the information.

Denial of Service is the prevention of authorized user from accessing a service or application by exhausting system or application resources.

Elevation of Privileges means that an ineligible user gets a privileged access to applications or services.

8.2.2   Data Flow Diagram

For analyzing an application using the STRIDE method, the architecture of the application needs to be represented by a Data Flow Diagram (DFD). In DFD, the application is decomposed into multiple components based on their functionality and then analyze the security threats for each component. Table 8.2 shows various DFD components, their representations and description in accordance to [22].

Additionally, each component is vulnerable to a group of threats that needs to be addressed. Data flows and data stores are vulnerable to tampering, information disclosure and DoS. Whereas interactors are vulnerable to spoofing and repudiation, Processes are vulnerable to all the threats.

Table 8.2 DFD components

Component	Representation	Description
Data flows	Arrow represents	Direction of flow of data
Data stores	Parallel horizontal lines	File database
Process	Circle	Application
Multi-process	Concentric circle	Compilation of various sub-process
Interactors	Rectangle	It represents end-points which provide as well as consume data in the system
Trust boundary	Straight line (Dotted)	It is the boundary between trusted and untrusted components

8.3 Security Analysis of B4 Application

Google WAN network is architecturally divided into two WANs. One network is used for directing users requests/responses and other named as B4, is used for synchronizing users data across geographically distributed data centers. The QoS requirements for data for both of these WANs are varied in nature, where the one which is directing users requests/responses, requires high bandwidth for data while the other one (B4) requires high availability for data. Furthermore, thousands of applications which run across B4 require a range of QoS classes. Some applications over B4 are low in volume and demand low latency while some are high in volume and demand high bandwidth [11]. The cost of maintenance of these varied number of QoS classes and an end-to-end application control motivated Google to implement the SDN technology.

Figure 8.2 shows the basic architecture of B4. It can be viewed as three logical layers i.e., Hardware, Controller, and Global. The hardware layer consists of commodity OpenFlow switches. The controller layer consists of a cluster of Network Control Systems (NCS) for fault tolerance, and an instance of Paxos. Paxos selects one NCS as a master for the site. Each NCS contains an OpenFlow controller (OFC) and Quagga. Quagga provides Border Gateway Protocol (BGP) connectivity between NCS server and Gateway and also exchanges network and traffic engineering updates between them. OFC hosts SDN application called Routing Application Proxy (RAP). RAP is subscribed to Quagga's traffic engineering updates and forward updates to OFC. Finally, the global layer consists of gateway and topology server. Whereas the gateway connects multiple data center sites, the topology server (TE) serves as the brain for the whole network and defines policies to engineer traffic between these sites.

8.3.1 Security Analysis

A general routing process of B4 is explained in Fig. 8.2. One way pointed lines denote the interface status change information communicated from the switch to the topology server and updated policies are sent back to the switch. (1) OpenFlow enabled switch sends the status information to the OpenFlow controller hosted on NCS, (2) The SDN application RAP forwards the information to Quagga, (3) Quagga using BGP update forwards the information to the Gateway, (4) Gateway forwards the information to the Traffic Engineering (TE) server, (5) TE having the full view of the network, constructs a tunnel for application, assigns a flow group to it, and forwards this information to Gateway, (6) Gateway uses the BGP updates to pass the information to the Quagga, (7) RAP using RPC receives Quagga updates, and forwards them to OFC. Using the receive updates, OFC updates its Network Information Database (NIB), (8) Using the received updates, OFC constructs new policies

1. Status change update
2. Updates sent to Quagga
3. Updates per site forward to Gateway
4. Update forward to TE
5. Policy generated
6. Forward to Quagga
7. Rap receive using RPC
8. Routing table updated

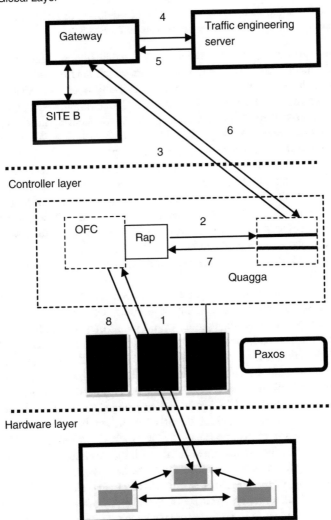

Fig. 8.2 B4 Architecture and flow of data [10] (redrawn and modified)

and forwards them to the OpenFlow switches. Finally, OpenFlow switches using new policies update their forwarding table.

For analyzing the security features of B4, we followed the steps mentioned in Sect. 8.2. For constructing a DFD considering the scenario depicted in Fig. 8.2, the application is basically decomposed into four main components—Topology server, NCS server, Gateway and OpenFlow hardware switches. Each NCS hosts the OFC, Quagga and two databases—Routing Information Database (RIB) and NIB. RAP application, Quagga and OFC are considered as DFD processes and two databases as DFD data stores. The data flows are specified according to the interaction of data with each DFD component in the graphical representation of DFD in Fig. 8.3.

A summary of the threat analysis is shown in Table 8.3.

Data flows   According to the STRIDE method, a data flow is vulnerable to tampering, information disclosure, and DoS threats which lead to violation of security goals such as integrity, confidentiality, and availability of data. Switches lie outside the trusted zone but use OpenFlow protocol for communicating with the OFC. OpenFlow uses Transport Layer Security (TLS) for securing the communication between the switches and controllers [4]. It protects the data flows from tampering of data and information disclosure by encrypting the data. It uses public key cryptography that ensures a private communication between the entities. Though controllers are kept in an isolated management network, to make the flows secure from the DoS attacks, filtering of OpenFlow requests or QoS of requests can be considered [22]. The data flow between the Gateway and NCS server and between Gateway and Topology server are vulnerable to all the three threats. Tampering and Information Disclosure can be mitigated by using encryption and message integrity check mechanisms, whereas DoS can be prevented by throttling mechanism or QoS.

Data store   Data stores are vulnerable to Tampering, Information Disclosure and Denial of Service attacks. In B4, data is stored in RIB and NIB. Compromising security of any of these data stores would result in vulnerability of the whole system. However, both the data stores are considered to be present locally inside the NCS Servers, so they can only be attacked if the NCS server is compromised.

Processes   Processes are vulnerable to all six threats categories, though spoofing, tampering and information disclosure could be neglected for them assuming that they are hosted on the NCS server and securing the server inherently secures them. To mitigate the repudiation threat, a log file can be maintained to store the details of each communication made to the process. For mitigating Elevation of Privileges attack, the process must be run with the minimum required privileges. For protecting from DoS attack, users must be authenticated and authorized and QoS or filtering of user request reaching to the process should be considered.

Interactors   Interactors are prone to spoofing and repudiation. In our DFD, switches, NCS Server, Gateway and TE server are the interactors. Switches and NCS servers may be considered to be safe from both of these attacks as they both use TLS enabled communication between them. Gateways and TE server can be spoofed if one is able to get access to the shell of an interactor. One can push certain

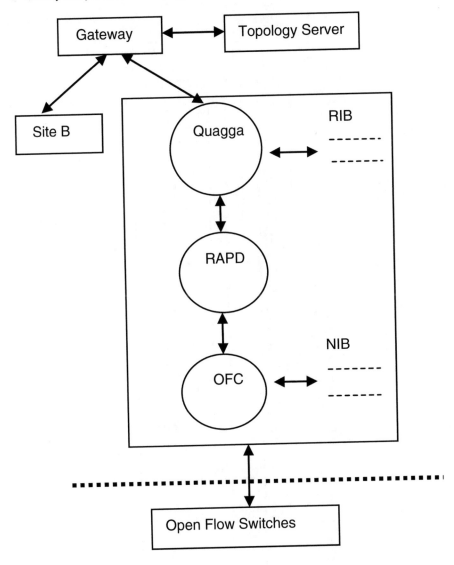

Fig. 8.3 B4 DFD

commands which can bring the application to a halt. In order to protect from such an attack, shell access must be authenticated and authorized. For example, usage of an AAA (Authentication, Authorization, and Accounting) server in the network could mitigate such attacks. AAA accounting feature will log each and every activity of a user, and any malicious attempt made by the user can be easily traced back, thus usage of AAA server mitigates repudiation attack as well.

Table 8.3 STRIDE matrix of B4 where "M" shows threat mitigation technique suggested in the paper, "*" indicates threat mitigated by the application architecture, "-" represents threats omitted, "G" represents gateway,"T" is the topology server, "N" is the NCS server, S means switches and "=" represents a two way data flow

Threats	Interactors				Processes			Data flows					Data Stores	
	G	T	N	S	OFC	RAP	Quagga	G = Quagga	OFC = S	OFC = NIB	OFC = RIB	G = T	NIB	RIB
Spoofing	M	M	*	*	*	*	*	-	-	-	-	-	-	-
Tampering	-	-	-	-	*	*	*	M	*	*	*	M	*	*
Repudiation	M	M	*	*	M	M	M		-	-	-	-	*	*
Information disclosure	-	-	-	-	*	*	*	M	*	*	*	M	*	*
Denial of service	-	-	-	-	M	M	M	M	M	*	*	M	*	*
Elevation of privileges	-	-	-	-	M	M	M	-	-	-	-	-	-	-

8.4 Security Analysis of IWAN Application

Cisco Intelligent WAN (IWAN), is one of the Cisco Application Policy Infrastructure Controller Enterprise Module (APIC-EM), application was developed for managing a hybrid WAN using SDN. It supplies all the capabilities of WAN management such as WAN optimization, performance routing, deep packet inspection and VPN tunneling in one suite. IWAN helps an organization to build Dynamic Multi-point VPNs (DMVPN) with zero touch deployment and intelligently route encrypted application traffic, independent of WAN transport. Additionally, it optimizes the WAN link using Cisco WAAS and secures communication using Cisco cloud security. It basically builds on four main components [23]: (1) Transport Independent Design, (2) Intelligent Path Control, (3) Application Optimization, and (4) Secure Connectivity.

8.4.1 Security Analysis

The functions of IWAN is explained using a basic scenario in Fig. 8.4. In the nutshell, headquarter (Hub) sends business critical traffic to private cloud using MPLS private link and during a certain time of the day the private link gets congested. Cisco IWAN observes the congestion and dynamically picks the low priority data from private link and re-route it to the other link such as Internet. For our study, we assume that the initial performance based routing policies for the application are already configured on the Border and Master router. The numbers in Fig. 8.4 represents the flow and it is described as follows: (1) The headquarter sends the application stream, (2) Master router uses the predefined policies to direct the traffic towards the Border router 1, (3) Border router 1 forwards the traffic to private Multi-protocol label switching (MPLS) link and using Next Generation Network-Based Application Recognition (NBAR2) starts deep inspection of the traffic, (4) Border router collects the metrics (Link utilization and Throughput) of application over private WAN link and using NetFlow V9 forwards them to the Master router, (5) IWAN receives metrics from the Master router, (6) As the traffic metrics reach threshold value, IWAN instructs new policies to the Master router and re-route the low priority applications running on private WAN link to the other WAN link (Internet), (7) Master router redirects the low priority data traffic to the Border router 2, 8) Finally, the Border router 2 forwards the traffic to the Internet link.

For the security analysis of IWAN, we followed the same steps as we did for B4. We first define the DFD (Fig. 8.5) and then analyze it using STRIDE. Basically, the architecture is decomposed into four main component groups—One or more LAN switch, Cisco APIC-EM controller, Master router, and both Border routers are considered as interactors for the DFD. The scenario in Fig. 8.4 reveals LAN and Master router, Master router and Border router, Border router and cloud, and Master router and IWAN communicate with each other. This communication is modeled as data flows. Furthermore, AVC, Performance Routing (PFR), and WAAS running

1.   Business traffic from LAN
2.   Master router Route using predefined policies
3.   Traffic passed to cloud (MPLS link)
4.   Net flow traffic metrics
5.   Metrics passed to IWAN
6.   New policies instruction
7.   Traffic redirected to other WAN (Internet Link)
8.   Traffic passed to cloud

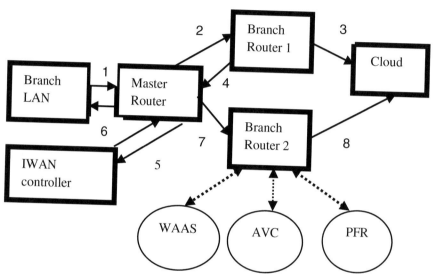

Fig. 8.4 IWAN architecture and flow of data [23] (redrawn)

on Border router and IWAN on APIC-EM controller are considered as processes. Finally, there are two trust boundaries. Branch switches are kept outside the trust boundary because users connected to the switches cannot be trusted, as they can overhaul the link between the Master router and Border routers with false traffic, which can hamper the IWAN performance. Second trust boundary is between edge router of the organization and WAN link. Data and users from the WAN link cannot be trusted either. There are no data stores considered for this application.

A summary of the IWAN security analysis is shown in Table 8.4.

Data flow   Data flow is vulnerable to Tampering, Information Disclosure, and Denial of Service [22]. Information Disclosure attack on the data flows between Master and Border routers can be performed by tapping the network between them. To achieve confidentiality, IPsec [24] tunnel between the routers can be implemented. Similarly, data can only be tampered if the attacker is able to tap

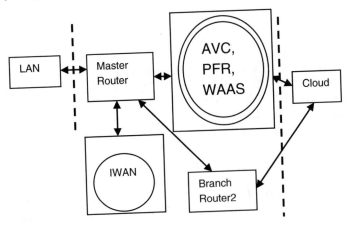

Fig. 8.5 IWAN DFD

the network and modifies the packets. This can also be mitigated by using IPsec tunnel which uses integrity checking mechanisms such as Message Authentication Codes (MAC). DoS attack on this flow could be performed if malicious users connected to branch LAN send false traffic and overload the link between the routers. To mitigate this threat, authentication and authorization of users connected to LAN and QoS of routing protocol and data metrics packets to be used. Data flows between APIC-EM controller and Master router are immune to both Tampering and Information Disclosure attack. Both the interactors use TLS enabled OpenFlow protocol between them [4] which makes the data flow immune to Tampering and Information Disclosure threats. Bandwidth of the link can be exhausted by sending large number of OpenFlow requests to the controller. To mitigate this, QoS or filtering of the OpenFlow requests can be considered [22].

Data flows between the branch LAN and the Master router are prone to all three attacks. Attacker with Tampering and Information Disclosure attack can manipulate the traffic, and by spoofing himself/herself, he/she can send bogus traffic on the WAN link, which can make IWAN to alter the performance routing configuration. This can be mitigated by using the TLS protocol between LAN and Master router and by authenticating and authorizing users on LAN. The flow of data from the Branch routers to the Internet is immune to Tampering and Information Disclosure attack because of IPSec IKE2 encryption [24]. Whereas, the DoS attack on the flow can surely hamper the working of IWAN. Unavailability to pass data through one link can cause IWAN to change the policies and routing entries on Master router, which can make the possibility of sending business critical data on the less secure Internet link. To mitigate this attack, one could either use dynamic IP on WAN ports of Branch routers or prioritize the business critical data on private WAN link.

Interactors   Interactors are prone to Spoofing and Repudiation attacks. In our DFD, Master router, Border router Controller and LAN switches are interactors.

Spoofing attack on the LAN switch can be performed if the attacker connected to switch, congests the WAN link with large amount of false traffic. This can make IWAN to alter the performance routing policies on the Master router. To mitigate this attack, we may either configure LAN switch ports to static access or disable Dynamic Trunking Protocol (DTP) auto-negotiation [25]. Repudiation attack on the LAN switch can be mitigated by logging the communication made to the switch using synchronized Syslog [26] and authenticating and authorizing the user connected to switch using AAA [27]. IWAN controller and Master router are both immune to Tampering and Repudiation attack because of secure OpenFlow TLS communication. On Border routers, Spoofing attack may be performed if console or management passwords are compromised. Cisco OS access is generally protected through AAA server which authenticates, authorizes, and accounts activity. For mitigating the Repudiation attacks, Border routers use DMVPN IPsec to build tunnels among the end-points. These tunnels use RSA signatures as one of the three options for the authentication of end points of the tunnel that provide immunity to Repudiation attacks [28].

Processes   As per STRIDE threat model, the process is vulnerable to all six threats. In our scenario, there are four processes WAAS, AVC, PFR, and IWAN application. Attacks will be easier to these processes if hardware hosting these processes get compromised. For example, if Border routers are compromised, then only the attacker will be able to execute attacks on AVC, PFR, WAAS. Spoofing and Tampering of processes can be performed by modifying the Cisco OS's (Cisco-Xe for PFR, AVC and WAAS and APIC-EM for IWAN) binary image. OS binary image can be modified by adding a malware to it. Cisco Image Verification can be used as a mitigation technique to this attack. This technique is built on the MD5 file validation and communicates any corruption in OS image to network administrators. Security of OS binary image can be further improved by providing authorization to commands like config-register and show memory only to certain users [29]. Similarly, Information Disclosure attack is possible if the attacker is able to access the binary image of OS (Cisco-Xe for routers and APIC-EM for the controller) and using the image the attacker performs static and dynamic code analysis. Code analysis can help an attacker to gain the knowledge of flow of data in the application (AVC, PFR, WAAS and IWAN). Using this knowledge, attacker can tamper the application data which can make an application to crash. Encrypting the binary file can be used as the mitigation technique to this type of attack. To perform DoS attack, an attacker connected to the branch LAN switch, can send large amount of data to router OS. These large amount of incoming request can crash the Router control plane and eventually shut down all the processes (AVC, PFR and WAAS). This attack can be mitigated by either using Cisco Copp [30] or by filtering the unwanted/malicious traffic incoming to the router's control plane. Elevation of privileges on all processes can be mitigated by running each process with the least privileges.

Table 8.4 STRIDE matrix of IWAN where "M" shows threat mitigation technique suggested in paper, "*" indicates threat mitigated by the application architecture, " - " represent threats omitted. "LAN" represent Branch, "Cont" is Controller, "M.R" is Master Router, "B.R" is Border Router and "=" represent two way flow

Threats	Interactors				Processes				Data flows			
	LAN	CONT	M.R	B.R	AVC	PFR	WAAS	IWAN	LAN = M.R	M.R = CONT	M.R = B.R	B.R = WAN
Spoofing	M	*	*	M	M	M	M	M	–	*	M	–
Tampering	–	–	–	–	M	M	M	M	M	*	M	*
Repudiation	M	*	*	*	M	M	M	M	M	*	M	*
Information disclosure	–	–	–	–	M	M	M	*	–	M	M	M
Denial of Service	–	–	–	–	M	M	M	M	–	M	M	M
Elevation of privileges	–	–	–	–	M	M	M	M	–	–	–	–

8.5 Conclusion

This paper presents the security analysis of two SDN-based WANs called B4 and IWAN. Both of these applications are widely deployed. The security analysis was done by using the threat modeling framework called STRIDE. Both of these applications use OpenFlow protocol for the communication between the controller with external hardware which is considered to be secure to Spoofing Tampering, Information Disclosure, and DoS attacks because of the presence of TLS. Before deploying these applications in the networks, the suggested security mechanisms can be considered.

References

1. The Zettabyte Era Trends and Analysis. http://www.cisco.com/c/en/us/solutions/collateral/serviceprovider/visual-networking-indexvni/VNIHyperconnectivityWP.html. Accessed 27 July 2016
2. Stallings W (2016) Software Dened Networks and OpenFlow. http://www.cisco.com/web/about/ac123/ac147/archivedissues/ipj16-1/161sdn.html. Accessed 27 July 2016
3. McKeown N, Anderson T, Balakrishnan H, Parulkar G, Peterson L, Rexford J, Shenker S, Turner J (2008) OpenFlow: enabling innovation in campus networks. ACM SIGCOMM Computput Commun Rev 38(2):69–74
4. Dix J (2016) The rst place to tackle SDN: in the WAN. http://www.networkworld.com/article/2873964/sdn/therst-place-to-tackle-sdn-inthe-wan.html?page=2. Accessed 02 Jan 2016
5. Software-dened networking: why we like it and how we are building on it. http://www.cisco.com/c/dam/enus/solutions/industries/docs/gov/cis13090sdnsledwhitepaper.pdf. Accessed 29 July 2016
6. Jacobs D (2016) How software-dened WAN architecture is changing the market. http://searchsdn.techtarget.com/tip/How-software-denedWAN-architecture-is-changinthemarket
7. Cisco Intelligent WAN. http://www.cisco.com/c/en/us/solutions/enterprise-networks/intelligent-WAN/index.html. Accessed 29 July 2016
8. Software-defined WAN: more uptime for your network, more downtime for you. http://www.cloudgenix.com/software-definedwan/. Accessed 29 July 2016
9. Secure Extensible Network (SEN) Solution. http://VIPtea.com/solutions/overview/. Accessed 29 July 2016
10. Jain S, Zhu M, Zolla J, Holzle U, Stuart S, Vahdat A, Kumar A, Mandal S, Ong J, Poutievski L, Singh A, Venkata S, WANderer J, Zhou J (2013) B4: Experience with a globally-deployed software defined WAN. In ACM SIGCOMM Computer Communication Review, ACM, vol 43, pp 3–14
11. Feamster N, Rexford J, Shenker S, Clark R, Hutchins R, Levin D, Bailey J (2013) SDX: a software-defined internet exchange. Open Networking Summit
12. Microsoft, improving web application security: threats and countermeasures. https://msdn.microsoft.com/en-us/library/ff648644.aspx. Accessed 02 Jan 2016
13. UcedaVelez T (2012) Real world threat modeling using the PASTA methodology. OWASP App Sec EU 2012
14. Saitta P, Larcom B, Eddington M (2005) Trike v.1 methodology document[draft]. http://dymaxion.org/trike/Trikepdf,2005
15. Schneier B (1999) Attack trees. Dr. Dobb's J
16. Jan Jürjens (2002) UMLsec: extending UML for secure SystEms development. In J ez equel JM, Hussmann H, Cook S (eds.) UML 2002, LNCS 2460. Springer, Heidelberg, pp 412–425

17. Christopher A, Audrey D, James S, Carol W (2003) Introduction to the OCTAVE ® approach. Carnegie Mellon University, Pittsburgh, pp 15213–3890
18. Misuse cases. http://ro.ecu.edu.au/cgi/viewcontent.cgi?article=1119contextism. Accessed 29 July 2016
19. Qualitative risk analysis with the DREAD model, Posted in General Security on 21 May 2014. http://resources.infosecinstitute.com/qualitative-risk-analysis-dread-model/. Accessed 29 July 2016
20. The CORAS approach to model-driven risk analysis. https://securitylab.disi.unitn.it/lib/exe/fetch.php. Accessed 29 July 2016
21. Johnstone MN (2010) Threat modelling with STRIDE and UML
22. Tasch M, Khondoker R, Marx R, Bayarou K (2014) Security analysis of security applications for soft-ware defined networks. In Proceedings of the AINTEC 2014, ACM, p 23
23. Cisco Intelligent WAN Design Guide. http://www.cisco.com/c/dam/en/us/td/docs/solutions/CVD/Jan2015/CVD-IWANDesignGuide-JAN15.pdf. Accessed 29 July 2016
24. An introduction to IP security (IPSec) encryption. http://www.cisco.com/c/en/us/support/docs/security-vpn/ipsec-negotiation-ike-protocols/16439-IPSECpart8.html. Accessed 29 July 2016
25. Dynamic trunking protocol. http://www.cisco.com/c/en/us/tech/lan-switching/dynamic-trunking-protocol-dtp/index.html. Accessed 29 July 2016
26. Cisco OS synchronized logs: Syslog. http://www.cisco.com/c/en/us/tech/ip/syslog/index.html. Accessed 29 July 2016
27. TACAS: accounting, authorization, acconting. http://www.cisco.com/c/en/us/support/docs/security-vpn/terminal-access-controller-access-control-system-tacacs-/10384-security.html. Accessed 29 July 2016
28. IPSEC RSA negotiation. http://www.cisco.com/c/en/us/td/docs/iosxml/ios/configuration/xe-3s/sec-ike-for-ipsec-vpns-xe-3s-book/sec-key-exch-ipsec.html. Accessed 29 July 2016
29. Cisco IOS security. http://www.cisco.com/c/en/us/about/security-center/integrity-assurance.html. Accessed 29 July 2016
30. Cisco control plane policy. http://www.cisco.com/c/en/us/about/security-center/copp-best-practices.html. Accessed 29 July 2016

Glossary

Authentication The process by which an entity, also called a principal, verifies that another entity is who or what it claims to be. A principal can be a user, some executable code, or a computer.

Authorisation Once a principal's identity is determined through authentication, the principal will usually want to access resources, such as printers and files. Authorization is determined by performing an access check to see whether the authenticated principal has access to the resource being requested.

Availability Readiness for correct service.

Big data Big data is data sets that are so voluminous and complex that traditional data processing application software are inadequate to deal with them. Big data challenges include capturing data, data storage, data analysis, search, sharing, transfer, visualization, querying, updating and information privacy.

Cloud networking Cloud networking (and Cloud based networking) is a term sdescribing the access of networking resources from a centralized third-party provider using Wide Area Networking (WAN) or Internet-based access technologies.

Confidentiality The absence of unauthorized disclosure of information.

Control plane The control plane is the part of a network that carries signaling traffic and is responsible for routing.

Data plane The data plane (sometimes known as the user plane, forwarding plane, carrier plane or bearer plane) is the part of a network that carries user traffic.

Date center A data center is a facility used to house computer systems and associated components, such as telecommunications and storage systems.

DoS A cyber-attack where the perpetrator seeks to make a machine or network resource unavailable to its intended users by temporarily or indefinitely disrupting services of a host connected to the Internet. In a distributed denial-of-service attack (DDoS attack), the incoming traffic flooding the victim originates from many different sources.

DPI Deep packet inspection (DPI, also called complete packet inspection and information extraction or IX) is a form of computer network packet filtering that examines the data part (and possibly also the header) of a packet as it passes an

129

© Springer International Publishing AG 2018
R. Khondoker (ed.), *SDN and NFV Security*, Lecture Notes in Networks
and Systems 30, https://doi.org/10.1007/978-3-319-71761-6

inspection point, searching for protocol non-compliance, viruses, spam, intrusions, or defined criteria to decide whether the packet may pass or if it needs to be routed to a different destination, or, for the purpose of collecting statistical information that functions at the Application layer of the OSI (Open Systems Interconnection model).

Firewall A gateway that limits access between networks in accordance with local security policy.

IDS A device or software that detects and notifies a user or enterprise of unauthorized or anomalous access to a network or computer system.

Internet of Things The Internet of things (IoT) is the network of physical devices, vehicles, home appliances and other items embedded with electronics, software, sensors, actuators, and network connectivity which enables these objects to connect and exchange data.

IPS A device or software used to prevent intruders from accessing systems from malicious or suspicious activity. This is contrast to an Intrusion Detection System (IDS), which merely detects and notifies.

Load balancer A load balancer is a device that acts as a reverse proxy and distributes network or application traffic across a number of servers.

Northbound API In a software-defined network (SDN) architecture, the northbound application program interfaces (APIs) are used to communicate between the SDN Controller and the services and applications running over the network. The northbound APIs can be used to facilitate innovation and enable efficient orchestration and automation of the network to align with the needs of different applications via SDN network programmability.

Repudiation Repudiation threats are associated with users who deny performing an action without other parties having any way to prove otherwise.

SDN Controller SDN Controllers (aka SDN Controller Platforms) in a software-defined network (SDN) are the "brains" of the network. It is the application that acts as a strategic control point in the SDN network, manage flow control to the switches/routers "below" (via southbound APIs) and the applications and business logic "above" (via northbound APIs) to deploy intelligent networks.

SDN The physical separation of the network control plane from the forwarding plane, and where a control plane controls several devices.

Southbound API In a software-defined network (SDN) architecture, southbound application program interfaces (APIs) (or SDN southbound APIs) are used to communicate between the SDN Controller and the switches and routers of the network. They can be open or proprietary.

Spoofing Spoofing threats allow an attacker to pose as another user or allow a rogue server to pose as a valid server.

Tampering Data tampering involves malicious modification of data.

Author Index

© Springer International Publishing AG 2018
R. Khondoker (ed.), *SDN and NFV Security*, Lecture Notes in Networks
and Systems 30, https://doi.org/10.1007/978-3-319-71761-6

Subject Index

133

© Springer International Publishing AG 2018
R. Khondoker (ed.), *SDN and NFV Security*, Lecture Notes in Networks
and Systems 30, https://doi.org/10.1007/978-3-319-71761-6

Printed in the United States
by Baker & Taylor Publisher Services